Ecological Imaginations in Latin American Fiction

UNIVERSITY PRESS OF FLORIDA

Florida A&M University, Tallahassee
Florida Atlantic University, Boca Raton
Florida Gulf Coast University, Ft. Myers
Florida International University, Miami
Florida State University, Tallahassee
New College of Florida, Sarasota
University of Central Florida, Orlando
University of Florida, Gainesville
University of North Florida, Jacksonville
University of South Florida, Tampa
University of West Florida, Pensacola

Ecological Imaginations in Latin American Fiction

Laura Barbas-Rhoden

University Press of Florida

Gainesville

Tallahassee

Tampa

Boca Raton

Pensacola

Orlando

Miami

Jacksonville

Ft. Myers

Sarasota

Printed in the United States of America. This book is printed on Glatfelter Natures
Book, a paper certified under the standards of the Forestry Stewardship Council (FSC).
It is a recycled stock that contains 30 percent post-consumer waste and is acid-free.

16 15 14 13 12 11   6 5 4 3 2 1

Library of Congress Cataloging-in-Publication Data
Barbas-Rhoden, Laura, 1974–
Ecological imaginations in Latin American fiction / Laura Barbas-Rhoden.
p. cm.
Includes bibliographical references and index.
ISBN 978-0-8130-3546-8 (alk. paper)
1. Latin American fiction—History and criticism. 2. Ecology in literature. I. Title.
PQ7082.N7B3155   2011
863.'30998—dc22
2010024733

The University Press of Florida is the scholarly publishing agency for the State
University System of Florida, comprising Florida A&M University, Florida Atlantic
University, Florida Gulf Coast University, Florida International University, Florida
State University, New College of Florida, University of Central Florida, University
of Florida, University of North Florida, University of South Florida, and University
of West Florida.

University Press of Florida
15 Northwest 15th Street
Gainesville, FL 32611-2079
http://www.upf.com

For John and Nick

Contents

Preface

This book has its academic origins in the Central American literary and cultural studies I have pursued as a professional. Truth be told, though, it had its real start much earlier. As a child, I divided my time between two passions: stories and the forests, fields, and swamps of the southeastern United States. The stories I heard at home from an immigrant father and Midwestern mother gave me an affection for faraway places I never knew intimately, like my parents did, and books taught me to appreciate many others. When I was not lost in stories, and sometimes with book in hand, I explored the bramble patches, creeks, and forests in and around the towns where my transplanted family lived. As an adult, I found academia and ecocriticism a comfortable fit, since both allowed me to explore multiple spaces and places with others. This project grew out of that early passion for the stories people tell and the places they inhabit, create, lose, leave, and imagine.

Though writing is a solitary enterprise, like most human endeavors, it would not happen apart from the collaboration of a community. I am fortunate to have a community of colleagues at Wofford College who have given me moral support and encouragement for this project. Among many to whom I am grateful, I owe a special thanks to nature writer John Lane for asking lots of questions and generously sharing with me his knowledge of North American nature writing and ecocriticism. I would also like to acknowledge the institutional support I received in the form of a faculty research grant, which allowed me to explore the feasibility of the project and funded my travel to the

Atlantic coast of Costa Rica in 2001. Additionally, I received two Community of Scholars summer faculty-student research awards, in 2006 and 2009, and wrote furiously as I mentored students. To my students in two classes on literature and the environment in Latin America, and to the Spanish majors who have followed this project with curiosity and interest, thank you for your eagerness to learn from others in the world you inhabit. Sharing it with you as you study is a privilege.

Others here and in places far from me have generously shared their time and knowledge. To my editor, Amy Gorelick, at the University Press of Florida, many thanks for beginning the conversation about this book at the 2004 LASA congress in Las Vegas and for your patience as it has unfolded. I am also grateful to Tatiana Lobo for stimulating conversations, in Costa Rica and by e-mail, as well as to Genny Ballard and Sofia Kearns for querying colleagues as I desperately sought scientific names for flora identified by colloquial monikers in novels. For assistance with the details of research, I am grateful to Paul Jones and the other staff members at the Sandor Teszler Library at Wofford College, as well as to the librarians of the Latin American Collection at Smathers Library at the University of Florida, where I conducted research in the summer of 2006. Other colleagues have been generous as readers. Julee Tate and Cissy Fowler read early versions of chapters and provided me with valuable critiques, as did the peer reviewers for *Mosaic* and *ISLE*, where I published preliminary versions of material included here. Chapter 1 appeared in abbreviated form in *Mosaic* 41.1 (2008), 1–18, and an ecocritical study of two novels from Central America appeared in *ISLE* 12.1 (2005), 1–17. I gratefully acknowledge the permission of those journals to reprint that material here. Two anonymous readers for the University Press of Florida reviewed the manuscript with extraordinary conscientiousness, and I am very grateful to them for their time and expertise. Of course, any errors and inaccuracies herein are my own.

Finally, I want to thank members of the closest community I have. Thanks to Autumn Harrell, for extra hours with my children, and to Anilda Varela, for being their unofficial *madrina* and for answering my endless queries about life in rural Honduras. To my parents, down in the wetlands of south Georgia, thanks for being my go-to source for science queries, stories, and endless moral support. A very special thank you to my husband Rob for balancing it all with me, and to my two sons, for making that balance careen wildly every day, even as you remind me to see the world with wonder and hope.

Introduction

"The message I want to send in a time capsule is a living message." Here Homero Aridjis pauses. "I would like to send a living monarch butterfly."

The Mexican poet speaks these words in the 2007 documentary *The 11th Hour*, and he names three creatures he would send into the future: the monarch butterfly (*Danaus plexippus*), gray whale (*Eschrictus robustus*), and sea turtle (several species from two families, *Cheloniidae* and *Dermochelyidae*). All are migratory animals that spend part of their lives in Mexico, but whose survival is as much linked to global forces as is Aridjis' own family story. Aridjis' poetic imagination sends them to flutter, splash, and nest in the imaginations of his global audience, motivating viewers to build a future that includes these creatures.

Homero Aridjis' participation in this documentary, narrated by Hollywood celebrity Leonardo DiCaprio, is part of an artistic movement to raise global consciousness about human and nonhuman nature. Aridjis is one of many committed public intellectuals in Latin America who have spoken to environmental realities. Harvard ecocritic Lawrence Buell states that "one sees a trend-line of mounting resolve during the late twentieth century to fathom and raise public consciousness about the history, present state, and possible future of the environmental interdependencies between human and nonhuman within human society, made increasingly unstable and dangerous by drastic alterations in planetary environment" (*Future* 97).

Aridjis certainly forms part of this global movement, but his art and activism also stem from a strong sense of place. The son of a Greek immigrant and a local woman, Aridjis grew up just miles from the wintering grounds of monarch butterflies in Michoacán, Mexico.

An ecocritical approach to Latin American literature can take many forms, but it must take into consideration the way literature registers both the intimate knowledge of place and global forces of change. Here, I am interested in the way contemporary Latin American fiction depicts the historical, economic, and cultural roots of ecological transformation and crisis in the Americas. It is my contention that texts of ecological imagination use a rhetoric of nature to expose and critique human power structures during a moment of growing unease about the global economy. The authors of the primary texts I consider all recuperate a sense of place, and some posit an ecocentric agenda that valorizes nonhuman nature. Others have an explicitly anthropocentric agenda, and their fictionalization of history and the natural world offers warnings about modernization and highlights paths of resistance to it.

I have selected texts with a dual purpose in mind: to offer a literary analysis of expressions of ecological imagination in contemporary Latin America and to produce an innovative text about the broad sweep of environmental history as apprehended through fiction. Specifically, I am interested in late twentieth-century novels of national or international significance that have an ecologically oriented imagination and that fictionalize moments in history the authors identify as determinant of ecological and social problems in Latin America. These works depict turning points in nineteenth- and twentieth-century environmental history, and they warn of an apocalyptic future for Latin America in the years to come. All also share a common preoccupation: the mobilization of a discourse of nature to indict the neoliberal order of late twentieth-century Latin America.[1]

The selected texts depict distinct time periods and cover four areas: Argentina at the turn of the nineteenth century, the Amazon in mid-twentieth century, Costa Rica in the late twentieth century, and Mesoamerica in a futuristic twenty-first century. As I have arranged them here, this broad selection of books becomes a vehicle to explore the environmental history of Latin America from south to north, past to present and future.

From Argentina, where nineteenth-century battles between civilization and barbarism played out famously in literature and the landscape, there are three novels: *Tierra del fuego* (1998) by Sylvia Iparraguirre, *Un piano en Bahía Desolación* (1994) by Libertad Demitrópulos, and *Fuegia* (1991) by Eduardo Belgrano Rawson. I examine three texts about the Amazon, by writers from three different countries: *Mad Maria* (1980) by Brazilian Márcio Souza, *Fordlandia* (1997) by Argentine Eduardo Sguiglia, and *Un viejo que leía novelas de amor* (1989) by Chilean-born Luis Sepúlveda. In Central America, I focus on literature from Costa Rica, the epicenter of the ecotourism industry. These primary texts are *Murámonos, Federico* (1973) by Joaquín Gutiérrez, *Calypso* (1996) by Tatiana Lobo, and *La loca de Gandoca* (1995) by Anacristina Rossi. In the last chapter, I look at two futuristic novels set in fictional Mesoamerican societies: *¿En quién piensas cuándo haces el amor?* (1995) by Mexican author Homero Aridjis and *Waslala* (1996) by Nicaraguan author Gioconda Belli.[2]

For this group of well-known contemporary Latin American writers, an ecological imagination does not mean a "nature-first" agenda. Environmental concerns in their fiction bolster a critique of contemporary economic and social realities. It is my contention that these books appropriate a rhetoric of nature and a preoccupation with heretofore marginal landscapes in order to question modernization in Latin America. Indeed, their very exploration of keystone moments of environmental history is "a sign of the sickness of the present," to borrow a phrase Jonathan Bate used in another context altogether (13).[3]

My analysis focuses on the way these Latin American texts appropriate the rhetoric of nature to push social and environmental justice concerns into the foreground of debates about history. In reading the novels, I practice a hybrid ecocriticism that draws on the work of historians and geographers, anthropologists and scientists. In terms of literary theory, I find narratology useful for considering the implications of the structure of the texts. Narratological tools like those elaborated by Gérard Genette and Wayne Booth provide the means by which to consider how narratives are constructed to convey a particular interpretation of history and, in some cases, to produce a moral effect on readers.

Postcolonial, Marxist, ecocritical, and ecofeminist theories also offer useful approaches for teasing out meaning. Reading strategies from postcolonial studies draw attention to the ways the novels portray the imperial designs that Great Britain, the United States, and Latin American political leaders have inscribed upon territories and communities. Marxist criticism adds a socioeconomic perspective that makes it possible to read the novels as narratives about the darker side of economic projects that cast shadows on both human beings and nonhuman nature. For example, the Argentine writers focus on the indigenous peoples and lands that are casualties of global designs, but they also bring into the foreground the thousands of metropolitan poor that empires mobilize to conquer lands, markets, and resources.

My project also aims to push along the debate about what a Latin American ecocriticism will look like, especially as it brushes against the forms the discussion has taken in European and North American literature. For me, Buell's reluctant conclusions to *The Future of Environmental Criticism* offer one viable prescription for ecocriticism for Latin America. Buell asserts that the soundest positions in humanistic environmental criticism "will be those that come closest to speaking *both* to humanity's most essential needs and to the state and fate of the earth and its nonhuman creatures independent of those needs, as well as to the balancing if not also the reconciliation of the two" (*Future* 127; emphasis in original). Latin American literature has a long tradition of registering the "most essential needs" of humanity, and it also demonstrates awareness that ecosystems in all regions are in increasing peril. The field of ecofeminism is useful, too, as it sheds light on discourses and praxes by which women and landscapes have been gendered and subjugated.

Ecocriticism on Latin American topics must take environmental, social, racial, and gender injustices into account. For the books included here, it is also imperative to incorporate historical perspectives on environmental topics. The authors levy a rhetorical challenge to forces of economic globalization and cultural homogenization, and in so doing, they make explicit connections between specific historical processes begun centuries ago and the violence that persists today against the natural world and the humans (particularly Indians, mestizos, migrants, rural communities, and other marginalized peoples) who inhabit it. For example, the texts use dark and mocking humor to retell stories of heroes of modernization, figures like Charles Darwin, Henry Ford, and Latin American elites. To counter the production and dissemination of "global designs" and the knowledge systems they superimposed on foreign territories, the texts focus attention on specific local knowledge systems and their connection to landscapes and ecosystems.[4] Finally, the texts articulate the need to retreat from modernization altogether or to challenge modernizing processes with a Latin American ingenuity that comes from dwelling deeply in place.

Ecocriticism and Literature from Latin America

Why apply ecocriticism to literature from Latin America? For ecocritics, many of whom have been working in Anglo-American and European traditions, creative works with an ecological conscience are important. Ecocritics read texts with the environment in mind, just as other scholars have read literature to bring the experiences of women, minorities, subalterns, and others into focus. Ecocritics believe that literature can offer communities of readers a deeper understanding of the world of nonhuman nature and the human place in it. Since ecocriticism includes a strong activist component, scholars in the field also value stories that shape imaginations and awareness of ecological realities. Many of us believe that art with an ecological imagination may ultimately shape the actions of a reading public, though prominent critics like Bate are quick to point out that "the business of literature is to work upon consciousness. The practical consequences of that work—social, environmental, political in the broadest sense—cannot be controlled or predicted" (23).

Much of the early history of ecocritical scholarship has been tied to Anglo-European literary traditions. In the United States, early ecocriticism focused on writers from the western United States, as well as the work of naturalists and intellectuals like John Muir and Henry David Thoreau. Given this past, Latin Americanists may be wary of the dangers of imposing theories developed in different contexts onto fields like Latin American literary and cultural studies. This is a valid concern, and theories and reading methodologies must be adapted, transformed, and sometimes discarded altogether when taken from their original contexts and applied elsewhere. However, contact and cross-fertilization of scholarship in literary studies is rewarding and necessary. Furthermore, cultural studies theorists like Antonio Candido and Néstor García Canclini have been making ecocritical points in their writing for some time without calling them so, and other scholars working in Latin American literary traditions have been engaged explicitly in ecocriticism, particularly over the course of the last decade.

For ecocriticism in general, and particularly as it is taught in the North American academy, there is benefit to be had in adding more voices to the already animated discussion in the field. When voices come from cultures where conceptualizations of nature and the relationship of literature to environment are different, these perspectives can challenge previously accepted categories of analysis. Though it has been a less prevalent practice for Latin Americanists than for others, there are also rewards for Latin Americanists when they read literature through an ecocritical filter. Questions of land, water, and resource use loom large over Latin American history. Literary scholars have analyzed Latin American literature on the basis of questions of aesthetics and with an eye toward themes like identity, gender, indigenous rights, social justice, memory, and so on. There are theoretical texts by Latin American scholars that articulate questions of identity, history, and knowledge systems. A growing Latin American–oriented ecocriticism can marry work in cultural studies to new theoretical elaborations from beyond and within Latin America to make a substantial contribution to the international conversation ongoing in ecocriticism.

Since much ecocriticism has been oriented toward texts from industrialized regions with long-standing environmental movements and different conceptions of "nature," certain concepts must be recognized by ecocritics as culturally specific. The idea of "nature" is one such concept. Early framers of ecocritical debates in the North American academy argued militantly for privileging nature writing and realism in literature. But "nature-first" agendas are incongruous in Latin America; they spring from particular, North American cultural traditions with roots in romanticism. Jennifer French points out that "in the tropical republics of the economic periphery, nature was the source of potential wealth and the site of economic growth and development" (13). Romanticism took a different form in a Latin American context, and so nature writing from nineteenth-century Spanish America is different: "Rather than the 'escape' from the realities of romanticism offered to European readers, Spanish American nature writing much more directly represents the continent's predominant economic forms and, as a result, its gradual incorporation into the international capitalist system" (French 13). The incorporation of Latin America into a global economy has had profound consequences for human beings and nonhuman nature during the last two centuries. For this reason, intellectuals from the region employ a discourse of nature in different historical moments to either endorse or resist the transformation of the region.

What Buell calls "second wave" ecocriticism is therefore a more logical place for a Latin American ecocriticism to engage with ecocritical debates, and indeed, that is happening. In second wave ecocriticism, scholars challenged the nature-first criteria for ecocentric literature. Buell himself adapted criteria he espoused in *The Environmental Imagination* and indicated that all variants of ecocentric thinking "define human identity not as free-standing but in terms of its relationship with the physical environment and/or non-human life forms" (*Future* 101). This looser definition of ecocentric thinking has evolved as the environmental justice movement challenges the terms of debate in ecocriticism.[5]

Because Latin American ecocriticism is in its incipient stages, it has the potential to model reconciliation of ecocentric and anthropocentric concerns in its praxes. In contextualizing the birth of ecocriticism among scholars involved in the Western Literature Association of the United States, Glen A. Love notes that "the late nineteenth- and twentieth-century West witnessed the transfer of the old machine-garden conflicts into the immediate present, with battles over the fate of the West's native peoples; over the appropriation of its water; over wilderness, old-growth forests, mineral extraction, endangered species, pollution, toxic wastes, and spreading urban blight" (*Practical* 4). Love writes about the American West, but the list could easily refer to Latin American realities in the twentieth and twenty-first centuries. The question, then, is how Latin American literature registers this march toward modernization, especially when it becomes accelerated by globalization at the end of the twentieth century. My book pulls together perspectives from cultural studies and ecocritical scholarship to challenge readers to recognize the cultural and economic dimensions of the discussion about literature and the environment, especially as it pertains to Latin America.

I argue specifically that in the late twentieth century, a growing number of Latin American authors produced texts preoccupied with natural spaces, and that they did so as part of a broader critique of economic systems of subjugation. Their discourse of nature represents a specific literary response to neoliberalism in Latin America. Anthropocentric concerns figure prominently in these texts, and because of this, they invite analysis along the lines suggested by the environmental justice movement. According to Joni Adamson, Mei Mei Evans, and Rachel Stein,

> environmental justice initiatives specifically attempt to redress the disproportionate incidence of environmental contamination in communities of the poor and/or communities of color, to secure for those affected the right to live unthreatened by the risks posed by environmental degradation and contamination, and to afford equal access to natural resources that sustain life and culture (4).

American cultural studies critic T. V. Reed adapts the definition of environmental justice to ask questions of literary relevance:

> How can literature and criticism further efforts of the environmental justice movement to bring attention to ways in which environmental degradation and hazards unequally affect poor people and people of color? How has racism domestically and internationally enabled greater environmental irresponsibility? . . . How can ecocriticism encourage justice and sustainable development in the so-called Third World? To what extent and in what ways have other ecocritical schools been ethnocentric and insensitive to race and class? (149)

By positing questions related to poverty, race, gender, and environmental degradation, the environmental justice perspective offers a useful global framework for discussion of literary texts.

Even in light of questions of global relevance, it is essential that ecocritics read Latin American texts with difference in mind. The Americas are a vast territory of myriad ecosystems and bioregions, from the Atacama and Sonoran deserts to the tropical forests of the Amazon and Central America. The cultures of Latin America are equally diverse: indigenous groups are numerous throughout the Americas, and they speak hundreds of different native languages. They have different pre-Columbian histories, as well as distinct experiences after European contact. Centuries of colonization and immigration introduced even more human communities, from African slaves to Russian Jews, and each brought their own practices of interacting with the nonhuman world to bear on Latin American landscapes.

Difference and diversity are essential in ecological terms as well as in cultural terms. According to ethnobotanist Gary Paul Nabhan, the occurrence of abundant biodiversity in regions of great cultural diversity is not a coincidence. He argues that "most biodiversity remaining on earth today occurs in areas where cultural diversity also persists. . . . It is fair to say that wherever many cultures have coexisted within the same region, biodiversity has also survived" (37). Some of the most biodiverse countries in the world are in Latin America, for example in nations such as Ecuador and Peru, where there is also great cultural diversity, especially among indigenous cultures.

Of course, Latin America is also a product of the spread of empire in the development of a modern world system. In the same region that astounds with dizzying varieties of life, one can see the negative corollary of Nabhan's earlier assertion: "wherever empires have spread to suppress other cultures' languages and land-tenure traditions, the loss of biodiversity has been dramatic" (37). The Spanish empire was just such an empire, one that brought large-scale mining, grain agriculture, and livestock grazing, with cataclysmic social and ecological transformations. Other great powers made further changes. First the British dominated trade in agricultural exports after Latin American independence, and then the United States held sway, consuming barrels of oil and minerals and tons of agricultural products each year for well over a century.

For ecocritics working in other literary traditions, it is imperative to understand that modern Latin America is a product of complicated webs of ethnic, social, and economic relationships. Born of centuries of miscegenation and cultural contact, its hybrid peoples possess myriad traditions and histories. There are people of mixed ethnicity, others of Spanish and Portuguese descent, as well as Afro-Latin Americans and more recent immigrants like the Chinese, Japanese, German, and Italian populations dispersed throughout the Americas. Different peoples have vastly different conceptions of their place in the world of nonhuman nature, and Latin America today is the site of their competing claims for the future.

Of course, companies, people, and organizations from abroad also have an interest in the natural wealth of Latin America, from mineral deposits in the Atacama Desert to tracts of arable acreage in Brazil. This is one reason that questions of land use dominate the history of Latin America. Many players inside and outside the region "read" Latin American people, the ecosystems they inhabit, and the stories cultures tell about these relationships. Where indigenous peoples see ancestral lands in Colombia and Ecuador, petroleum companies see energy reserves and profits. Rural people in Honduras look out on farmlands; mining companies see gold. Tourists in Costa Rica find entertainment in coastal ecosystems rich in aquatic and avian life during an afternoon of snorkeling. Conservation organizations see parks and reserves in Brazil where governments see potential agricultural land, migrants see a chance at a better life, and indigenous groups see opportunity in ecotourism or cattle ranching.

Latin American authors think, write, and publish in this milieu. Though many are not themselves deeply attached to one place (by choice or by exile), their works depict the intricacies of national and local histories, cultures, and landscapes. Moreover, artists and intellectuals from the Americas often draw on a vast network of personal contacts, published reports, oral histories, literary works, and their own imaginations to craft their stories. Most published authors occupy positions of relative privilege that allow them to engage with people from diverse disciplines, social strata, and professions within national borders and beyond. For this reason, the Latin American authors I consider here read the human and nonhuman world with one foot planted in global culture and another in the regions that have captivated them. When ecocritics read these Latin American stories (from Latin America or from a desk in a university in the United States), they tap into different, and usually multiple, conceptions of the natural world and the place of humans in it in the past and future.

Debates about the future of Latin America abound in nearly every corner where people concern themselves with issues such as biodiversity, poverty, democracy, or development. Increasingly in environmental circles, debates take into consideration the fate of humans and ecosystems together, a focus long missing from debates among conservationists in the developed world. Until the 1990s, U.S. and European conservation organizations mostly ignored the socioeconomic realities of populations in the vicinity of their projects. Timothy O'Riordan, editor of *Environment*, wrote an article "On Justice, Sustainability, and Democracy," which was published in the journal in 2005. In it, he explains that "at the most principled levels, organizations advocating for human rights, social justice, and nature's intrinsic value have not yet managed a coordinated meeting of the minds" (0+). Partly in response to critiques such as these, environmental organizations based in the

United States and Europe increasingly work with social justice organizations to bridge the chasm between efforts in conservation and social justice. They are also making efforts to educate their supporters about the need to address human as well as nonhuman concerns in order to ensure successful, sustainable conservation efforts. For example, *Nature Conservancy* magazine featured the cover article "The Poverty/Conservation Equation," in Summer 2006.[6]

The human/nonhuman dichotomy has begun to be addressed in Anglo-European environmental circles, but it is debatable whether it has ever existed in the same way in Latin America. Pressing human concerns have dominated progressive social agendas among students and activists in Latin American universities and intellectual circles, from the 1968 student protests in Mexico City to demonstrations against the Free Trade Agreement of the Americas in Mar del Plata, Argentina, in 2005. Human concerns are intertwined in environmental struggles, as in the case of the Awá in Colombia, whose lives and territories have been caught in a protracted and bloody conflict between government forces and rebel groups. Robust indigenous movements, like that of Ecuador, have grown in strength and numbers since the 1990s, and they, too, have offered different perspectives on how human and nonhuman nature exist in the Americas.

When authors from Latin America express both social justice and ecological concerns in literary texts, the way writers present this convergence is instructive. For instance, the novels included here all draw on history to promote an agenda of environmental and social justice. They express strong misgivings about the economic order of the world, and they historicize the marginalization of minority groups and the exploitation of ecological wealth. They associate the rhetoric of empire and nation-building with a political drive toward cultural homogeneity and the accumulation of wealth on the part of Latin American elites and foreign investors. Some texts, like those of Iparraguirre and Lobo, may surprise readers outside of Latin American letters with their skepticism of science and scientists, since both authors note the entanglement of science with imperialism and gently mock its founding fathers. Yet not all is critique when it comes to scientific knowledge; the texts also contain scientifically correct, local knowledge of the nonhuman natural world. An additional contribution is their sense of the more-than-human world, a dimension with deep spiritual significance, often rooted in indigenous and folk traditions, in which animal and botanical life takes on supernatural significance. Authors interweave references to scientific facts and the more-than-human world in lyrical and elegiac passages, thereby disrupting facile Cartesian binaries that have held sway for centuries in Western thought about nature.

These literary texts with an ecological awareness also feature an ethical message intended to shape the imaginations of readers, many of whom are likely urbanites in Latin America and the rest of the world. Just as earlier traditions brought other social justice struggles into focus, from the social realism of Upton Sinclair in the United States to the consciousness-raising *testimonios* against dictatorships in Central and South America, so, too, does this literature of ecological imagination seek to guide readers toward a more complex understanding of human history in the landscapes of Latin America.

Latin American Literary History and the Place of Ecocriticsm

In terms of literary history, the works I explore are post-Boom novels, texts that inherit and defy the legacy of what has been called the Latin American "Boom" of the 1960s and 1970s. But what came before the Boom and post-Boom? How do these contemporary texts figure in Latin American literary history, and what can be the role of ecocriticism in reconsidering Latin American literary history? Here I would like to give a brief overview of Latin American literary history and also highlight ecocritical perspectives on literature from earlier periods. To do this, I draw heavily from an essay on literary history by Brazilian critic Antonio Candido. His comments explicitly link the arts to conceptions of the natural world, and they are a useful place to start a discussion of ecocriticism in Latin America.

In an essay originally published in 1970, Candido commented on the literary history of Latin America since independence.[7] His essay surveys early movements like romanticism, regionalism, and modernism, tracing literary history all the way through to literary movements in vogue when he published his essay. Candido himself draws on comments from fellow Brazilian writer Mário Vieira de Mello to argue that until the 1930s, Latin America was still in the grip of the idea of the "new country," a reality unrealized but full of promise. Candido asserts that, in the hands of Latin American intellectuals in the decades immediately after independence, "literature became the language of celebration and tender reflection . . . with support from hyperbole and the transformation of exoticism into a state of the soul" (36). In this moment, "the idea of *country* was closely linked to that *of nature* and in part drew its justification from it" (36; emphasis in original).

After independence, literary movements often represented nature in a positive, sometimes euphoric frame, and nature was seen as fundamentally important to national development. Jennifer French points out that "the foundational discourses of Spanish America are for good reason dominated by a rhetoric of nature: nature was to be the economic basis of the new republics, in the eyes of both European capitalists and the creole elites who shaped their post-independence development" (13). Later, when post-independence euphoria gave way to what Candido calls the "consciousness of underdevelopment" that followed the Second World War, the nature metaphor persisted, but it now registered a more pessimistic consciousness of the position of Latin American in economics and politics (Candido 37; French 15).

The literature of a consciousness of underdevelopment shifts focus to social, technological, and cultural problems, and "the resulting vision is pessimistic with respect to the present and problematic with respect to the future" (Candido 37). In this vein occurred expressions such as naturalism and social realism, indigenism, and the novel of the northeast of Brazil (Candido 54). Later in the twentieth century, Candido argues, works with antinaturalist techniques appeared in "a blooming world of the novel" (54). This was the literary scene at the time he published the essay. Later referred to as "the Boom," the works that begin to be published in the late 1960s are, according to Candido, "nourished by nonrealist elements, such as the absurd, the magic of situations, or by antinaturalist techniques, such as the interior monologue, the simultaneous vision, the synthesis, the ellipsis" (56). Despite these technical innovations, though, Candido argues that the novels continue to explore "the very substance of nativism, of exoticism, and of social documentary" (56).

The Boom and, ironically, globalization, opened the way for writing from many quarters in Latin America at the end of the twentieth century. The post-Boom period brings proliferation: the rise of testimonial literature linked to human rights struggles; literature penned by women, indigenous authors, and Afro-Latin Americans; urban crime novels; and post-dictatorial fiction. Many critiques of modernization that García Canclini sees in contemporary anthropology also mark fiction from the last decades of the twentieth century. García Canclini has argued that contemporary anthropologists are "critics of modernity" who have rejected "evolutionism and ethnocentrism" ("Cultural Studies" 332).[8] Their oppositional posture "induces them to understand the homogenizing policies of industrialization and industrial reconversion, of national integration and subordination to transnational patterns of development, either as Western impositions on ethnic and local cultures and of the hegemonic classes on the subaltern ones, or, in the most radical cases, as simple ethnocide" (332). The same could be said of many Latin American writers of the same period.

For example, nearly all the texts I include here ultimately posit a retreat from modernization as part of their narrative trajectory. At the same time, this contemporary fiction from Latin America is itself a product of the modernization it critiques: beneficiary, product, and contributor to a global exchange of ideas. Even the interest in history might be attributable to modernity; Jonathan Bate holds that "the demand for historical explanations as well as, or instead of, mythical ones is one of the characteristics of 'modernity'" (29).

The contemporary novels of ecological imagination that I have included here have another common thread that reveals a preoccupation with modernization. They are the product of the clash and combination of cultures, and nearly all feature protagonists or major characters who are outsiders or hybrid figures. They are mestizos, nonindigenous peoples, and alienated, displaced people; they are people who have been relocated or who have rejected the cultures that were their birthrights. As a result of personal crises or geopolitical forces, these characters move into new territories where they come into contact with Others. Positioned between cultures and places, they have a different perspective on the ecological and social clashes they witness. Like the authors who constructed them, these protagonists perceive and communicate competing human conceptions of landscape and place. By means of these figures, the novels contrast the embeddedness and rootedness of certain populations, juxtaposing this against the rapacious nomadism of exploiters and affirming the testimonial function of the writer and dissident. The key interpretive figure is the protagonist who learns to read the social, political, and ecological landscape of Latin America and grapple with the crises she or he sees.

How much are these topics and characters a reflection of the anxieties and ambitions of the authors? Most Latin American authors have lived nearly all their lives in urban centers and have traveled widely within Latin America and beyond its boundaries. Many experienced displacement and exile for political reasons. Because of their education, mobility, and personal circumstances, the writers are well positioned to convey their understanding of Latin America to national and global audiences. They are also nearly unanimous in their critique of the global economic system. The authors of ecologically oriented texts make an additional point, arguing that modernization has been predicated on the exploitation of natural resources and on the marginalization or elimination of other modes of living on the planet.

Overview of Chapters

The emergence of works of ecological imagination and critique coincides with a surge in the pressures of economic globalization in Latin America. Buell holds that "nature has been doubly otherized in modern thought," stating that empirical reality serves human interests and one of those interests has been to reinforce the subservience of marginalized groups (*Environmental Imagination* 20). Ecologically conscious Latin American authors reclaim the value of both the human and the more-than-human in a gesture that contests this "otherization." Whether it fully valorizes nonhuman nature, or just repeats the otherization of nature in the vindication of the human Other is another question—perhaps even another book.

In laying out an ecocritical reading of these texts, my intention is to give readers new to the field or region a sense of the possibilities for ecocritical analysis in contemporary Latin American letters. Since my study limits itself to narrative texts that make a particular environmental critique of neoliberal modernization and that show the progression in time of patterns of ecological destruction in Latin America, many voices that deserve close ecocritical attention are not included. I cite a few examples here in hopes that their mention inspires further scholarship. Contemporary authors from the Movimiento Maya, many of whom implicitly or explicitly address the legacy of genocide in the construction of the modern state, are not here. Authors like

Rigoberta Menchú and Gaspar Pedro González engage debates of modernity in ways that are radically different from the authors I include, with literary interventions that are more often about the right to a place and space in the nation-state, rather than about the infringement of globalization upon places important to Mayan communities. Studies have been done on another area I omit—the linguistically and historically diverse Caribbean, but more work is warranted about the way this literature registers a common Caribbean environmental history.[9] Finally, I reference Afro-Latin American oral traditions, but I have not included narratives by authors who self-identify as part of Afro-Latin cultures, and certainly the work of authors like Quince Duncan, who speaks about the exclusion of black Costa Ricans from the nation-state, merits ecocritical analysis.

A final word is due on the organization of chapters. I have organized chapters by region and by the historical periods the novels consider, rather than by date of publication or national origin of the author. The reason for this is twofold. First, the story of Latin American environmental history needs to be told, just as Buell calls for it to be told in the United States (*Future* 115). An examination of selected fiction about different historical moments, from early republic to modern failed states and megalopolises, is one way to tell this environmental history. Second, ecocritic Scott Slovic argues for "contextualization and synthesis" on the part of ecocritics who want to do meaningful work with their literary scholarship. Approaching literary works by region and by the historical period they depict allows me to tell a compelling story of environmental history, while providing contextualization and synthesis for a body of works with a great ecological imagination.

My first chapter takes as its subject a group of twentieth-century Argentine texts that examine the realities of Latin America after independence. This was a critical period when leaders pursuing civilizing agendas pushed up against their own heterogeneous populations and the powerful British empire. The second chapter is on fictions written about the Amazon. My reading of these novels brings into the foreground the social and economic repercussions of extractive economies and infrastructure construction over the course of the twentieth century. In Chapter 3, Costa Rican texts pick up the story of Latin American environmental history in the mid- to late twentieth century as the republics of the isthmus embraced agricultural export economies and neoliberal policies associated with globalization. Finally, the last chapter on futuristic fiction considers two texts that draw from the rich history of Latin America to condemn environmental degradation, political corruption, and the disintegration of human communities in the present. My hope is that this organizational scheme will make key moments in Latin American environmental history, as represented in contemporary fiction, accessible to readers interested in national literary traditions, historical fiction, and environmental concerns.

Chapter 1

Alterity, Empire, and Nation in Tierra del Fuego

Fabled for its breathtaking landscapes and windswept seas, Tierra del Fuego lies far from the metropolitan centers of the world. Despite its physical distance from London and even Buenos Aires, Tierra del Fuego became intertwined in the complex, global economic system that has taken shape over the last two centuries. The history of these southernmost lands is a tale of the transformation of landscapes and cultures in the march to modernization. Contemporary stories about this change are a promising introduction to the ecological imaginations of Latin American authors.

Tierra del fuego (1998) by Sylvia Iparraguirre, *Un piano en Bahía Desolación* (1994) by Libertad Demitrópulos, and *Fuegia* (1991) by Eduardo Belgrano Rawson all take the people and place of Tierra del Fuego as the focus of their stories. Penned in the 1990s, each novel fictionalizes the same transformative moment in Argentine and world history in the nineteenth century. Specifically, the texts depict cultural conflict and environmental change in the critical nineteenth and early twentieth centuries. The novels reveal a confluence of ecological, historical, and political concerns that are central characteristics in ecological imaginations in contemporary fiction.

In their novels about Tierra del Fuego, the writers chart the transformation of a hitherto marginal landscape as it became subsumed into the evolving world economic system. In particular, they speak to the presence of the ecological and cultural diversity in Argentina that fell to an advancing capitalist frontier in the nineteenth century. By focusing on indigenous cultures in Tierra del Fuego, the novels make explicit connections between human and environmental casualties of the new world order inaugurated by European colonialism and Latin American nation formation.

The novels return to a foundational moment in national history to critique present-day social, economic, and environmental structures that are a product of that historical moment. In doing so, they engage a nineteenth-century historical and literary discourse that occluded the subservience of Latin America to the British empire. As Jennifer French argues, foundational literary texts from the nineteenth century, such as Andrés Bello's *La agricultura en la zona tórrida* (1826), establish two patterns that continue into the present. The first is the "almost complete occlusion of neo-colonialism in Spanish American cultural discourse in the nineteenth century" (6). The second is the displacement of the tensions created by neocolonialism onto a discourse on nature: "Britain menaces Spain by threatening to usurp its colonies, which are implicitly identified with nature" (7). By taking on the foundational moments of the nineteenth century, the novels dismantle the official history of nation formation and link the cultural and ecological problems it established to the neocolonial world order of globalization in which the novels were written. The ending of each novel seals the message of the story; each posits either a retreat from modernity or a determined confrontation of it.

Contextualizing the Novels

Iparraguirre, Demitrópulos, and Belgrano Rawson are well-known writers with distinguished careers in Argentine letters. Long active in Argentine literary circles, Iparraguirre won several prizes for *Tierra del fuego*, including the prestigious Sor Juana Inés de la Cruz prize in 2000. In that same year, *Tierra del fuego* was translated into English and published by Curbstone Press. Demitrópulos is likewise highly regarded, having published numerous works before her death in 1998.[1] Likewise, Eduardo Belgrano Rawson is author of several novels and recipient of literary accolades in Argentina. Within months of publication in Argentina, *Fuegia* was a critical and editorial success and was already in its fourth edition (Matthieu 147).

The territory of Tierra del Fuego that all three authors consider is far removed from the urban world of Buenos Aires, the preferred setting for many contemporary Argentine novels. The novels shun the urban landscape to consider one even more remote than that preferred by regional writers of the 1920s and 1930s. In turning a literary eye toward Tierra del Fuego, the authors draw attention to a part of Argentine territory central to the emergence of the modern world system and to the sailors, scientists, and indigenous people who traversed the lands and canals.

Tierra del Fuego represents the southernmost extension of Argentine territory. It is a landscape of pampas, mountains, canals, islands, and ice floes. Tierra del Fuego Island itself has three distinct zones inhabited by different cultural groups. The upper part of Tierra del Fuego resembles the lands in Patagonia to the north, with flat, open grasslands; it was the territory of the P'amica Ona people (D. Wilson 115). The southern part of the island is marked by plains and high mountains, with forest cover in parts; this was the territory of the Hámska Ona (D. Wilson 115). The southeastern zone of Tierra del Fuego Island was the territory of the Aush people, who subsisted mostly on marine mammals (D. Wilson 116). South of Tierra del Fuego Island is an archipelago that was inhabited by the Yahgan-Yámana canoe people; the Chilean archipelago to the west was the territory of the Alacaluf (D. Wilson 129). Belgrano Rawson traversed the Mitre Peninsula, the easternmost part of Tierra del Fuego Island, with a team of Argentine biologists, and his novel is a byproduct of that expedition.

For centuries after Columbus' voyages, indigenous groups inhabited Tierra del Fuego with only intermittent encounters with Europeans. The Spanish made an early attempt at establishing a presence in the area in 1584 under the leadership of Sarmiento de Gamboa. The troubled colony survived only a brief time, however, as colonists died in the harsh conditions. Supplies and reinforcements never arrived, and only a handful of people lived to tell the story of the settlement.

There are few descriptions of the indigenous peoples of the archipelago until the early twentieth century. This was after the nineteenth and twentieth centuries had delivered sweeping change to Tierra del Fuego. The rapid transformation of the world economy at the end of the nineteenth century coincided with the establishment of new Latin American states. It was thus a critical time for the environment and peoples of the Americas. The incorporation of Tierra del Fuego into the modern world system involved a catastrophic transformation of lands and cultures and the erosion of their attendant, place-based knowledge systems. In such a way, the inexorable march of empires and nations changed landscapes that indigenous groups had inhabited with relative stability for centuries (D. Wilson 120).

In Tierra del Fuego, as in other places, a sustainable subsistence economy organized around the rhythms of a particular place changed in a few decades into an extractive economy in which natural resources like land, minerals, and petroleum were controlled, exploited, and depleted. Indigenous populations declined through disease, violence, and the exhaustion of the resources upon which their lives depended. With the demise and eventual disappearance of indigenous communities, elements of the landscape lost their cultural significance, as well as the protections local knowledge systems provided to prevent over-exploitation of the nonhuman natural world. New mentalities and patterns of consumption displaced local knowledge systems that once provided protection against resource depletion. Land, flora, and fauna became commodities, part of a massive transfer of wealth for consumption in Europe. Patterns of consumption changed in Argentina, too, as newly arrived (mostly European) settlers boosted the total population and national elites profiting from an export economy conspicuously consumed products, from British soaps to metropolitan opera, in vogue in Europe (French 23).

The ecologically minded fictions of Iparraguirre, Demitrópulos, and Belgrano Rawson interweave dramas of the human and more-than-human worlds to render decades of radical change in new ways. With the benefit of hindsight, their novels reinterpret the impact of British colonialism in Latin America. The "Invisible Empire," as French calls the British presence in Latin America, becomes visible (15). This is in large part accomplished through an appropriation of a nature-oriented rhetoric that draws on historical documents, as well as indigenist and regionalist traditions in Latin American letters.

Within the context of Argentine literary production, the texts are noteworthy because they consider the diversity of cultures in the Southern Cone. They question a national identity founded on an affinity for European influence and promulgated by elites from metropolitan centers like Buenos Aires. By depicting areas at the fringe of the national imaginary, the novels restore to memory places that exist far beyond the highly touted urban spaces and beyond the heavily mythologized pampas. Indeed, their focus brings to light peoples and places "disappeared" in Argentine history long before the more famous disappearances of dissidents during the Dirty War.[2] In this case, the authors' exploration of the recovery of memory, a theme recurrent in post-dictatorial literature, takes readers deep into national and world economic history.

For ecocritical studies, the novels are significant because they make artistic interventions in debates about people and places in the modern world. The novels question modern realities in part by resituating important actors in the story of modernization. For example, the novels included in this chapter re-create the moment when modern science emerged around figures like Darwin and other naturalists. In retelling the story of naturalists, physicians, and anthropologists from the perspective of the global South, the authors bring the imperial connections and ethnocentrism of science back into focus. Scientific knowledge came into being through collaborations with empire, and it overlapped in uncomfortable ways with the campaigns of British missionaries and economic speculators.

In a contemporary context of great anxiety about globalization in Latin America, the novels return attention to the foundational moments of Latin American economies in order to question their supremacy. The novels push the boundaries of the imagined community of the nation, transporting readers to landscapes altered by the foundation of Latin American republics and the export economies their governments promoted. To express their profound unease about these, the authors employ a rhetoric of nature that draws from a long tradition in Latin American literature.

In crafting new historical fiction about Tierra del Fuego, Iparraguirre, Demitrópulos, and Belgrano Rawson invite readers to see "Western Civilization" and the story of its elaboration from below, from the global South. The rise of modern colonialism, particularly British naval and commercial supremacy, had profound implications for environmental change in a newly independent Argentina. Because the novels delve into this history, a short exploration of the context will be helpful for the ecocritical analysis that follows.

As northern European powers challenged the old empires of Spain and Portugal, the colonial era began for many more regions of the world in the eighteenth and nineteenth centuries. The original colonial era in Latin America is different; it began shortly after the 1492 voyage of Columbus when Spain first established settlements in the Caribbean. At the height of the Spanish empire, Spanish territorial claims (though not always political or military control) extended southward from western North America to Tierra del Fuego. Spanish imperial control endured three centuries, until the early nineteenth century when Latin American territories began declaring independence and waging long wars to make the declarations a reality.[3] Mexico, for example, declared independence with the Grito de Dolores in 1810, but Spain did not recognize Mexican independence until 1821. By the mid-1820s, all but the Caribbean possessions of Spain (and the Philippines) had gained their independence.

Despite the political and military victories of these wars, independence in Latin America did not produce the same social and political upheaval as did independence movements in Africa and the Middle East a century later. In fact, as prominent Peruvian sociologist Aníbal Quijano argues, the end of the colonial era and the "metamorphosis of modernity" in Latin America "served to excessively prolong a system of power whose beneficiaries were social groups that embodied the most perverse results of colonial domination" ("Modernity" 148). For the most part, racial and social hierarchies remained undisturbed after independence and, Quijano asserts, became even more deeply entrenched.

Furthermore, Latin American independence coincided with the emergence of a power-grabbing modernity in Europe ("Modernity" 148). Latin America was no longer a subject of empire in the nineteenth century, but its peoples and places felt the effects of colonialism in profound ways. In fact, the Latin American position relative to Northern European imperial powers in the decades immediately following independence explains a great deal about the way Latin America articulated into the new world economy.[4]

The nonhuman natural world of Latin America changed in profound ways as governments acted on liberal principles of progress, order, and trade. In collaboration with foreign enterprises, local elites in government instituted policies designed to exploit land and labor for foreign exchange and the accumulation of wealth. For Latin American authors with ecological imaginations, the history of the transformation of labor, land, and politics is crucial to understanding the environmental and social challenges that face the region today. The contemporary ecological imagination critiques an economic order it perceives to have begun after independence. By drawing attention to erased histories and marginalized landscapes, it unmasks the rhetoric that has promoted the expansion of the capitalist frontier and justified the violent incorporation of humans and nonhuman nature into modern systems of trade.

Tierra del fuego, Un piano en Bahía Desolación, and *Fuegia* depict changes wrought by nineteenth-century imperial designs as foundational in establishing practices and attitudes toward cultural and biological diversity in the Americas.[5] The novels undertake the important task, as Latin American philosopher Enrique Dussel defines it, of "denying the innocence of modernity and of affirming the alterity of the other" ("Europe" 473). For the authors of these texts, the physical and cultural transformation of Latin American landscapes during the era of colonialism is one that sets the course for future societies and their attendant crises. According to the novels, the modern world system coming to dominance in the nineteenth century was a space of cultural and commercial encounters with profound consequences for ecosystems and communities.

As a network of global trade, the modern world system had been set in motion at least since the discovery of the Americas, if not before.[6] By the nineteenth century, this global economic model had already produced profound effects for human and nonhuman populations in the Americas. With the industrialization of Europe, just when countries such as Argentina achieved their independence, further changes were in store for the Americas. Britain sought cheaper raw materials and new markets for its manufactured goods in order to edge out European rivals and maintain a balance of trade with Asian powers, particularly China.

Iparraguirre, Demitrópulos, and Belgrano Rawson rewrite history to highlight the cultural and ecological crises catalyzed by British imperial projects during the critical years of nation formation in Latin America. Among European initiatives were programs to chart and survey territories using new technology, to catalogue and collect flora and fauna, and to expand land in production for agricultural commodities (especially those used in the booming textile industry). Latin American elites also wanted to expand agricultural production for export-led growth and domestic consumption. This economic program, coupled with the political need to assert control over borders, pushed settler populations into more and more territory (Bulmer-Thomas 46). Through liberal legislation and, in Argentina, through immigration, populations began to move into the interior and peripheries of the country, to places like the vast pampas and Tierra del Fuego.

The political rhetoric used to frame the expansion was the discourse of civilization versus barbarism. The name most closely associated with this rhetoric in Latin America is that of Domingo Faustino Sarmiento, author of *Facundo* (1845), also known as *Civilización y barbarie*. A founding figure for Argentine national identity, Sarmiento "can be deified or vilified but certainly never set aside" (Goodrich 1). Sarmiento frames the debate about the future of Argentina in binary terms: it is a battle of city versus country, educated men versus indigenous and gaucho savages, civilization versus barbarism. For Argentina to progress as a nation, according to Sarmiento, it must fill its empty lands, educate its citizens, and import European immigrants and technological progress. Consider this famous passage from the opening lines of the text:

El mal que aqueja a la República Argentina es la extensión; el desierto lo rodea por todas partes, se le insinúa en las entrañas; la soledad, el despoblado sin una habitación humana, son por lo general los límites incuestionables entre unas y otras provincias. Allí, la inmensidad por todas partes; inmensa la llanura, inmensos los bosques, inmensos los ríos, el horizonte siempre incierto, siempre confundiéndose con la tierra entre celajes y vapores tenues que no dejan en la lejana perspectiva señalar el punto en que el munco acaba y principia el cielo. Al Sur y al Norte acéchanla los salvajes, que aguardan las noches de luna para caer, cual enjambre de hienas, sobre los ganados que pacen en los campos y en las indefensas poblaciones (22–23).

[The disease from which the Argentine Republic suffers is its own expanse: the desert wilderness surrounds it on all sides and insinuates into its bowels; solitude, a barren land with no human habitation, in general are the unquestionable borders between one province and another. There, immensity is everywhere: immense plains, immense forests, immense rivers, the horizon always unclear, always confused with the earth amid swift-moving clouds and tenuous mists, which do not allow the point where the world ends and the sky begins to be marked in a far-off perspective. To the south and the north, savages lurk, waiting for moonlit nights to descend, like a pack of hyenas, on the herds that graze the countryside, and on defenseless settlements (45–46).]

The verbs especially are indicative of attitude. Argentina suffers from immensity; desert surrounds islands of civilization; mists confuse the boundary between earth and sky; and savages lurk like bloodthirsty animals. Tierra del Fuego does not figure per se in Sarmiento's opening chapter, but the Argentine national program of expansion and the British capital investments that fueled it both impinged greatly upon the austral territory.

Before the nineteenth century, European interest in Tierra del Fuego had been minimal. Nevertheless, the waterways of the region served as an important transportation route in the modern world system from the time of Magellan until the construction of the Panama Canal. By the nineteenth century, ships engaged in the business of empire frequented the area, and Tierra del Fuego became a space of cultural encounters, some of them violent. The British were among the first Europeans to chart the territories, leading them to several decades of contact with native inhabitants.

By midcentury, the British had established a permanent presence on the Malvinas, or Falkland Islands. A *New York Times* article refers to a small settlement there with "a sufficient number of soldiers to assist the authorities in maintaining the Government" and notes that the "design of nature in the location of these islands appears to have been to provide a convenient stopping-place for vessels in their long and wearisome voyage round Cape Horn" ("Sea and Ship News"). By 1860, in recounting the murder of British missionaries in Tierra del Fuego, the same newspaper reported that "all the political machinery of a colony" could be found on the "cluster of islands to the eastward of Tierra del Fuego" ("Murder of Missionaries"). The same article conspicuously praised the abundant water, livestock, peat fuel, and wild fowl in the area.

As the language of the article reveals, Britain had a clear geopolitical interest in the region. By the 1880s, both Argentina and Chile had grown sufficiently concerned about the British presence to assert sovereignty in Tierra del Fuego. In order to defend their territorial claims in the face of the British presence and to establish a lasting peace between themselves, the two Latin American nations with claims in the region agreed to divide Tierra del Fuego with a north-south line that partitioned the territory into east and west (British claims continued in the islands known as the Falklands or Malvinas). According to this agreement, neither Chile nor Argentina would have access to both oceans, something the rivalrous nations feared would tilt the balance of power.

By the time of Chilean-Argentine diplomatic agreements, though, Tierra del Fuego had already long been a contested territory. Smugglers, seal hunters, missionaries, native peoples, and ship captains rounding the Cape, all moved along its shores in search of livelihood. With national consolidation efforts in Latin America, the fate of its human and nonhuman inhabitants changed rapidly and irrevocably in a matter of decades.

In the national schemes of nineteenth-century leaders in Argentina in particular, outlying regions like Tierra del Fuego and Patagonia took on added importance. Elite nation-builders worked to fix the foundations of society upon an export economy tied to Great Britain by bonds of trade and debt. At the same time, the desire among Latin American elites to adopt European models for arts and society was strong. According to Argentine scholar and historian Walter Mignolo, "nineteenth-century intellectuals of the Southern Cone politically embraced the civilizing mission . . . while modern technology (the frontiers, the railroad) was being exported to the Southern Cone and was part of the emergence of new colonialisms (Britain, France) and the fading away of the old ones (Spain, Portugal)" (201). Economics and science emerged as powerful disciplines and discourses that went hand in hand with nation-building (and with the empires against which new nations sought to defend themselves). Liberal policies of nineteenth-century Latin American politicians advocated an export-led economy, infrastructure development, European immigration, and expansion into frontier areas.[8] All this meant a dramatic transformation of landscapes and peoples in areas like Tierra del Fuego.

In a new era of globalization in the 1990s, the novels of Iparraguirre, Demitrópulos, and Belgrano Rawson return to this early, formative era, and they appropriate a rhetoric of nature to critique it. Each novel deals with a slightly different time period in Tierra del Fuego. Read together, they retell the history of an era and promote a different vision of national history. According to their version of the past, the programmatic conversion of territories bolstered the imperial designs of Great Britain and promoted the national consolidation plans invented by elites. Argentina entered the world economy, but the cost to people and ecosystems was great.

Just how were territories altered, though? One example is illustrative of the monumental change. Plans for export-led growth encouraged the conversion of lands for new uses. Patagonia and Tierra del Fuego, for example, saw the enclosure of lands and the establishment of the sheep industry. Much of this industry was controlled by barons who were themselves recent immigrants to the Americas. The importation of sheep and immigrants quickly precipitated a campaign against the guanaco (an alpaca-like grazing animal). Native to the region, the guanaco was vital to indigenous peoples as a source of fibers for clothing and protein for consumption. Like the guanacos, native peoples were also an impediment to the prosperity of the sheep industry, so they were pursued and killed by bounty hunters on behalf of sheep barons who perceived both as obstacles to progress. Thus, economic changes (such as the growth of the export economy for wool and hides) led to the elimination of ecologically and culturally diverse populations. Government programs (like the "Conquista del Desierto" that settled Patagonia and squelched indigenous resistance), along with the actions of powerful individuals like the sheepowners, obliterated human groups deemed culturally undesirable and drove nonhuman populations into decline.

Why, though, do the authors consider this time period now, more than one hundred years after the trauma of change? I argue that they do so in order to insert the story of nineteenth-century conflicts and changes into a present-day conversation about national history and neoliberal economic policy. The appropriation of the discourse of nature is fundamental in their task. These particular post-dictatorial novels reconstruct history to recast confrontations of a political or economic nature as disputes about the fundamentals of human existence: how people live in the nonhuman world and how they reckon their lives—in ideology and in action—for themselves and others.

Tierra del fuego

Iparraguirre's *Tierra del fuego* deals with an early period of exploration by the British in Tierra del Fuego. The diegesis, or narrated story, unfolds during the time period from the 1830s to the 1860s. Iparraguirre also uses flashbacks, or analepsis, framed in the memories of particular characters, to reach even further back in history. Most published criticism of the novel has focused on this manipulation of history by Iparraguirre. Only a handful of articles have been published in all, though, and only one (my own) considers the ecocritical implications of Iparraguirre's narrative.

Tierra del fuego is rooted in history, but it is also rooted in place. The plot revolves around Jack Guevara, the bastard son of an Englishman and a creole mother from the upper classes of Buenos Aires. Guevara is born on the pampas in the midst of the Argentine struggle for independence. His unmarried parents retreat there, and his mother dies when he is quite young. After his father commits suicide, the teenage Guevara makes his way to the port and spends the better part of his life as a sailor on British merchant ships. He eventually returns to the pampas after curiosity, awe, and youthful energy have been sated by decades of travel.

At the start of the novel, Guevara has long since retired from the sea and is living a solitary life on the pampas. One day a written request from a British navy official interrupts his respite. The letter requests an explanation of an event that took place twenty years earlier involving a Fuegian native known as Jemmy Button. The novel details the lives of Guevara and Button, and the discourse moves between two moments: that of the exploits of young Guevara and his later reflection on that time. Passages in which Guevara comments on his own process of composition interrupt the flow of the narrative about his earlier years, intermittently reminding the reader of the circumstances of the construction of the history he is telling.

At the center of the plot is a historical episode, an actual event well-documented by nineteenth-century sources. This is a keystone event for Iparraguirre because it reveals the dramatic changes to the human and nonhuman worlds of Tierra del Fuego at a moment of imperial and national expansion. The "Button episode" began when young British sea captain Robert FitzRoy and a group of Yámana Indians met in waters off the coast of Tierra del Fuego in the 1830s. An aristocrat and naval prodigy in his twenties, FitzRoy was the new commander of a vessel surveying the coasts of Tierra del Fuego (the years of the voyage were 1826–1830) for the British Admiralty. During the course of the journey, FitzRoy took four young Yámana Indians from their communities in Tierra del Fuego. According to FitzRoy, this act was retaliation for the theft by native Fuegians of a valuable whaleboat attached to his ship, and he hoped the hostage situation would induce the Fuegians to produce the boat. The whaleboat never materialized, and FitzRoy transported the four Fuegians to England to be educated at FitzRoy's own expense. A favorite of the group, according to both the novel and historical accounts, was a young indigenous man the captain named Jemmy Button.

After some time in England, where the group of Fuegians enjoyed celebrity status, FitzRoy returned Button (and the other two surviving Yámana) to Tierra del Fuego. They sailed again on the *Beagle*, the Admiralty vessel most famous for Button's shipmate on that return voyage, the young naturalist Charles Darwin.[9] FitzRoy's intention was for Button to evangelize his people and act as a friendly contact for British military and commercial interests in Tierra del Fuego. This did not happen; in fact, Button was implicated in the murder of British missionaries decades later. In her novel, Iparraguirre explores why, probing also the significance of Button's life story for his fictional Argentine contemporary.

In *Tierra del fuego*, Iparraguirre recasts the actual historical events that surrounded the life of Jemmy Button and Charles Darwin into a unique narrative about empire and its impact on Jemmy Button's culture. She achieves this principally by inserting a fictional Argentine sailor into the FitzRoy's crew. This interloper shares the name of one of Argentina's most famous citizens, Ernesto "Che" Guevara, the revolutionary leader killed in 1968 in Bolivia. Like Che, Iparraguirre's Guevara straddles two worlds, one of privilege and another of marginality, and the trajectory of his life tends toward a self-identification with people on the margins of Latin American history.

Prior to a recent spate of writings, the events in which Button was implicated appeared only as curious anecdotes in the imperial history in the United Kingdom.[10] For instance, in 1973, a text that Everest mountaineer Eric Shipton published on Tierra del Fuego briefly described Button's adventure in naïve terms. Consider this excerpt about Button's return to Tierra del Fuego, the last phrase of which Shipton quotes verbatim from FitzRoy's memoir of the voyage:

> [A] few hours later [they] reached an spot known as Woollya, in Jemmy's country, where they selected a site for the settlement. It was an idyllic place, particularly in the warm sunshine. Rising gently from the shore there was a wide area of rich pasture, gay with flowers and watered by clear streams. This was sheltered by low hills covered with woods of the "finest timber trees in the country" (99–100).

From the description, the landscape could be Sussex or Kent, and the writer revels in a moment of "warm sunshine" in a region more famous for frigid, windswept landscapes. The imperial gaze, nostalgic in the 1970s, still looks for what it wants to see: green grass for grazing, fertile soil for crops, and happy, civilized natives tending fields and cutting fine timber.

In contrast to the above passage, Iparraguirre's narrative marks difference. Tierra del Fuego is not Kent; it is a collection of islands inhabited by peoples with a unique knowledge of their place in a fragile ecosystem. Her novel appropriates the Button episode to tell an Argentine story. In contrast to earlier authors, from Darwin to travel writers, Iparraguirre draws a sharp focus on the subjectivity of the human beings who inhabit these islands and understand their unique natural history. Furthermore, her novel communicates that the Fuegian clash with colonial powers was as much about environment and place as it was about identity and autonomy.

Iparraguirre's novel, like other Latin American econarratives, constructs a fiction that imagines and valorizes the interconnectedness of cultural diversity and biodiversity. Diversity flourishes together, she asserts, and it falls before the assault of totalizing, imperial systems. In *Tierra del fuego*, the will to cultural and commercial domination on the part of Great Britain is met with great resistance. Button challenges Guevara's efforts to objectify him, and his subjectivity changes Guevara's life. Similary, Iparraguirre shows the agency of Button's people, who are not just victims of imperial projects in his ancestral home, but individuals who actively negotiate and resist the encroachment that eventually obliterates them.

According to cultural studies scholars like Walter Mignolo, imperial projects negated the logic of indigenous knowledge, while at the same time using overwhelming force to break the foundations of indigenous life.[11] British armies, bureaucracies, and missionaries around the globe imposed a social order linked to the peculiar local histories of Britain. In *Tierra del fuego*, a passage that juxtaposes the teaching of Fuegian elders and the indoctrination of British manners in Jemmy Button during his time on a country farm underscores the contrast between cultures. In the episode, Guevara visits Jemmy on the farm in England, and Button tells him of his deep desire to return home. He also asks Guevara about the teaching of Guevara's own elders. Guevara tells him he learned from his father to break horses, hunt, and sail; Button shakes his head at the list, and the annoyed Guevara asks about Button's own beliefs. It becomes clear that Button is asking about place and values:

Aprende a renunciar a todo exceso. Todos y cada uno, sea hombre o mujer, deben mostrar el mayor respeto por los ancianos. Ancianos saben cómo construir el wigwam y la canoa, cómo luchar con la ballena, te ayudarán a vivir, te consolarán, te contarán de antepasados (149). [Learn to give up all excess. Each and every one of us, man or woman, must show greatest respect for old people. Old people know how to build the wigwam and the canoe, how to fight against the whale, they will help you live, will console you, will tell you about ancestors (103).]

Iparraguirre's text goes on to reveal the extent to which the British appropriated the place Button called home for their own projects. In tapping its resources for economic gain, they not only destroyed Button's people but also dismantled the knowledge and ethics they elaborated in their native place.

For Iparraguirre, the result of imperial processes is that the knowledge of entire populations is discredited and destroyed. This has been a point argued by Latin American philosophers and sociologists, as well. As Dussel points out, "The exclusion, as a civilizing criterion, of everything non-European . . . gave Europe—which already had military, economic, and political hegemony—cultural and ideological domination" ("World-System" 232). Through her mobilization of a rhetoric of nature, Iparraguirre adds another dimension that Dussel overlooks in his critique: Europe reorganized space in Latin America. The imposition of hegemony was played out in specific places.

Reading again the landscape upon which the British resettled Button (that pastoral site with fields and distant timber), it is reasonable for readers educated in European sensibilities to appreciate why such a place might be selected and later idealized in the captain's memoir. Iparraguirre thus selects this episode to signal cultural differences in reading landscapes. For instance, her narrative points out that the Fuegians themselves would never have picked such a site. The openness and view valued by a pastoral aesthetic would leave the Fuegians vulnerable to attack by other indigenous peoples and outsiders. The homogenization of societies and places throughout the Americas came precisely from the predilection of European settlers for superimposing their own aesthetic and economic preferences on landscapes they themselves did not know intimately. With the expansion of modern empires, places and peoples began to fulfill new functions. The actors in a world economy thus relegated local knowledge about place to the margins of their own projects, ones that revolved around the accumulation of wealth and power rather than around the rhythms of the natural world.

Iparraguirre structures her narrative in such a way as to emphasize that the capitalist conquest that swept over people and territories in the nineteenth century was a failure for all individual human beings directly involved in it. In general terms, the advancing capitalist frontier wrought havoc on "cultures of habitat" (to borrow Nabhan's term) like that of Jemmy Button, and it led to widespread assaults on ecosystems. Iparraguirre points also to specific human and environmental costs. The experiment with Button culminated in the murder of several English missionaries years after his return to Tierra del Fuego. Though the missionaries did not perceive themselves as such, the Yámana Indians interpreted them as the vanguard of invaders, in this case the British empire. For the Yámana, the missionaries were the pioneers who opened the way for military and commercial incursions that destroyed the teaching of the Yámana elders. Under the alleged leadership of Button himself, the Yámana killed the intruders, and the last pages of the Iparraguirre novel feature the trial of Button, which Guevara attends and observes.

According to *Tierra del fuego*, the imperial project failed the Yámana because it negated their agency, knowledge of place, and desire for self-determination. She depicts the Yámana as people who refused to play assigned roles in an imperial drama. The Yámana did not welcome passing ships nor did they rescue shipwrecked sailors. Rather they assessed the situation like any other individual and gamed it (by taking gifts of food, for example) as long as they could. They drew the line at becoming an outpost of agrarian stability and British gentility. Those who rebelled by pursuing their own modes of existence dwindled from sickness and the depletion of the resources upon which their lives depended.

The account of the death of Button's wife depicts the convergence of exterminating forces on Button's family and the place they inhabit. In this episode, Iparraguirre draws attention to the gendered violence located at the intersection of the patriarchal privilege and environmental destruction in imperial projects. In Yámana society, women traditionally gathered food, while men hunted. Because the encroaching newcomers have not respected norms such as the prohibition against killing baby animals, a tenet observed by Fuegian cultures, resources have quickly become depleted. As a consequence, native women increased the range over which they fished and collected seaweed and crustaceans.

> El año anterior había sido de desgracias. Un invierno feroz trajo el hambre como hacía mucho tiempo no se padecía en las islas. Una de sus mujeres, o su mujer anterior, no entendí bien, había sido violada y asesinada por loberos, precisamente por arriesgarse demasiado en busca de comida (204–5).
>
> [The previous year had been disastrous. A fierce winter had brought such hunger as had not been suffered on the islands for quite a long time. One of his wives—or his previous wife, I didn't quite understand—had been raped and murdered by seal hunters precisely for taking too big a risk in hunting for food (143–44).]

To the seal hunters, Button's wife is an unaccompanied (thus unprotected and therefore sexually available) native woman. The passage, narrated in detail, is the centerpiece of the last conversation Guevara has with Jemmy in Button's homeland. They will not meet again until Button's trial in the Falkland Islands. With the murder of Button's wife, Iparraguirre draws attention to the way ecological and gender concerns converge in injustice and tragedy. She also portrays the later murder of missionaries as an act the Yámana perceived as self-defense.

Though Button is a central, tragic figure, Iparraguirre catalogues casualties of modernization from many classes. FitzRoy commits suicide, as did his uncle before him. Jemmy Button dies on his islands, but not before he has witnessed the beginning of the extinction of his people and the plunder of the island habitat they guarded. Guevara never feels a part of the glorious enterprise of civilization (either in Britain or Argentina), and he retreats to the pampas.

In the person of Guevara, Iparraguirre invents a vehicle by which to critique empire from a unique perspective and also to posit an alternative model of existence for Latin America.[12] As the son of an immigrant and a Creole, Guevara is an Argentine "everyman," born at roughly the same time as the nation. However, Guevara feels the personal rejection of the creole elite as a youth and is later deeply troubled by personal encounters with the barbarism of empire. As a consequence, he makes a conscious retreat from the colonial enterprise. This withdrawal to a simplicity of life on the pampas is not, however, a vindication of the pastoral. Rather, Iparraguirre presents it as a logical and ethical response to the trauma of colonization, one that takes the protagonist to an archetypal Argentine place associated with independent thought, ingenuity, and resilience.

In the novel, Guevara's residence in the pampas eventually becomes the physical place and metaphysical space from which he can critique the modernizing project. Disquieted by the request from the British navy, Guevara writes to reconcile his own history, place, and identity. Initially troubled by the request, he later becomes suspicious: "Inglaterra rara vez se ha preocupado desinteresadamente por el mundo que existe detrás de los hechos. Y lo que ha terminado por interesarme es justamente eso: lo que hubo detrás de los hechos" (61–62) [But England has hardly ever cared disinterestedly about the world existing behind the facts. And what I have ended up being interested in is precisely that: what was behind the facts (36)]. He explores "what was behind the facts," writing his account in order to come to terms with his own complicity in a project about which he had profound misgivings. Guevara becomes a lens through which Iparraguirre filters history and models alternative ways of understanding the past and present. Guevara has observed the Other observing him, and he internalizes some of the Other's perspective on himself and his own culture.

With this novel, Iparraguirre challenges the imperial gaze on landscape and peoples.[13] She thereby advocates a new Latin American understanding of people and place. For example, in one passage, Guevara responds to "míster MacDowell or MacDowness" (the writing on the British missive is barely decipherable) and asks if he has ever seen a Fuegian, "ese ser extraño," "un hombre igual que usted, que inventa dioses, que caza y hace la guerra y domestica perros y enciende el fuego" (64) [that strange being . . . a man like you and me who . . . invents gods, hunts, goes to war, domesticates dogs, and starts a fire (38)]. Guevara declares that his experiential knowledge matters; behind Button, he found a people as complex as he, but with a spirit and a respect for life he has never encountered elsewhere (64).

The personal transformation of Guevara is striking because Iparraguirre explicitly links cultural coming-to-consciousness to growing ecological awareness. As a young man, Guevara internalizes other sailors' fears about Cape Horn and the austral lands. He likewise disdains Button as culturally inferior, and as a result, he disregards the landscape. Off the coast of Tierra del Fuego, before meeting Button, Guevara sees only terrifying weather and a graveyard of ships; in short, the tropes of a "liquid hell" straight out of the novels his father read (88). However, Guevara's contact with Button and the Yámana changes the way he perceives land and sea. Eventually Guevara observes that "el largo trato de los años con ese lugar de veranos apacibles y paz sobrenatural pudo reemplazar aquella primera impresión" (89) [Only the long familiarity of years with that place of mild summers and supernatural peace was able to change that first impression (56)]. With his perception corrected by contact with autochthonous cultures, Guevara gives up the tropes of imperial rhetoric and grapples with articulating a new vision of place.

The portion of the novel dedicated to the return voyage of the *Beagle* serves to underline the complexity of Guevara's native interlocutor and to return the imperial gaze of the European, in this case, naturalist Charles Darwin. Long passages detail the famous disagreement between the religious fundamentalism of FitzRoy and the scientific rigor of Charles Darwin. But Iparraguirre also firmly places Darwin in his own cultural context, carefully pointing out the ethnocentrism of the famous scientist. Consider this dialogue between Darwin and Guevara:

—¿Cómo? —me decía, haciéndose el incrédulo—. ¿Es que los gaúchos [sic] saben leer? Cómo es posible el prodigio, si son salvajes. [sic] Yo me reía, pero por algún lado me picaba. (164)
["What?" he would say to me, looking incredulous, "Do gauchos know how to read? How is this miracle possible, if they're savages?" I laughed, but it stung me in a way (114–15).]

Guevara likes the young scientist, but he bristles at the reflection of his own Otherness in Darwin's eyes. Button, for his part, limits himself to mocking Darwin's seasickness and observing in silence the fiery debates between the scientist and the captain.

After Button returns to Tierra del Fuego, Guevara sees him only briefly over the course of the next three decades. It is eventually Guevara's appreciation for Button as a human being, coupled with his own knowledge of Argentine places, that distances him from a European (in this case, British) vision of the world. Because Guevara understands what British imperial rhetoric despises, he becomes aware of the epistemological limits of European supremacy. The close knowledge of landscapes and peoples ultimately gives Guevara the confidence to defy his British interlocutor. Consider this passage:

> Una enorme porción de tierra patagónica aparece en esos viejos mapas bajo la denominación res nullis, cosa de nadie. Es mi país. ¿Ha estado usted alguna vez en la Patagonia, mister MacDowell o MacDowness? ¿Puede siquiera imaginársela? . . . ¿Puede imaginar esta inmensidad donde caben mil Londres? (89)
> [An enormous portion of Patagonian territory appears on those old maps under the name res nullis, no-man's land. It's my country. Have you ever been in Patagonia, Mr. MacDowell or MacDowness? Can you even imagine it? . . . Can you imagine an immensity which could hold a thousand Londons? (57)]

The passage echoes the opening of Sarmiento's famous *Facundo*, but Iparraguirre crafts a tone that is proprietary and proud in the voice of Guevara. Landscape, culture, and meaningful contact with another person teaches Guevara to recognize the shortcomings of a powerful empire and its representatives. Gradually aware of his subaltern status in their world, he responds confidently, with an altered sense of self and place.

Through Guevara, then, Iparraguirre not only explores competing representations of the human and nonhuman world of Tierra del Fuego, she also postulates an avenue for resistance. Guevara rejects the European representations he had initially accepted with complicity and a sense of awe. By the end of the novel, Guevara has written a folio of pages on the intersection of his life with that of Jemmy Button, and he has documented his personal transformation as a result of that contact.

Iparraguirre's protagonist allies himself with nature against empire and republic. As he writes to process his own history, he identifies with Button, Tierra del Fuego, and the pampas. Consider this passage, in which Iparraguirre makes clear that her novel's critique is levied against the new global system in which the British play a central role:

El mundo que conoció Button, el de sus antepasados, empezaba su largo fin. Como los témpanos desprendidos de los ventisqueros, su mundo comenzaba a requebrajarse y pronto navegaría a la deriva rumbo a su propia disolución. Mi situación no era demasiado distinta. Ibamos al encuentro de un orden que no preveía otro lugar para nosotros salvo el que ya nos tenía asignado. Veníamos de los bordes del mundo, de los confines de un lugar insospechado y bárbaro, que a pesar de mi buen inglés y mi crencha rubia emanaba de mí y me rodeaba, de igual manera que rodeaba a Button (103–4).

[The world Button knew, that of his ancestors, was coming to its long end. Like icebergs broken loose from glaciers, his world was beginning to disintegrate and would soon drift toward its own dissolution. My own situation was not very different. We came from the outer edges of the world, from its ultimate limits, from a barbaric, unimagined place which, in spite of my good English and my blond mop, emanated from me and surrounded me, just as it surrounded Button (67).]

Latin America is Other, as Guevara is Other. The implication is that, like him, Latin America should accept its alterity and write its history and future for itself.

In an act of defiance against the prerogatives claimed by Great Britain upon his memory, Guevara decides not to send his missive to the British authorities who requested it.[14] Instead, he looks for a different reader for the history and finds one in the domestic servant Graciana who lives with him. Over the course of the narrative, Guevara has become increasingly aware of Graciana as a person. Throughout the novel, Graciana is a silent presence, a domestic servant who shares Guevara's home and occasionally his bed. As he writes, Guevara becomes more aware of her as a person. Passages change from minimal commentary to careful observation, then register collaboration.

In the last pages of the novel, in an act rich in symbolism, Guevara decides to teach illiterate Graciana to read and write: "Si éste es un relato para nadie, quizá yo mismo deba crearle un lector, y tal vez sea ella" (284–85) [If this is a story for no one, perhaps I should create a reader for it, and perhaps it is she (199)]. With that decision, Guevara widens the conversation about history and place. Where nature and woman-as-nature were once only background in Guevara's drama, they are now in the foreground of the next stage of his life. Graciana will be empowered to do what she wants with the troubled history of her master, her country, her world.

Un piano en Bahía Desolación

Un piano en Bahía Desolación by Libertad Demitrópulos also retells the story of the British empire in the Tierra del Fuego archipelago in the nineteenth century. The action takes place in the mid-to-late Victorian era, a few decades later than the events in the Iparraguirre novel, in and around Punta Arenas and Bahía Desolación, on the Chilean side of the archipelago. Read together with *Tierra del fuego*, the novel offers a glimpse into the next stages of ecological and cultural change. Though it uses different narrative techniques, it, too, posits a critique and retreat from modernity that stems from contact with Latin America's Others and the places they inhabit.

As with Iparraguirre, there is remarkably little published criticism of Demitrópulos' work. Most of what has been published concerns another fictional text, *Río de las congojas* (1981). Much like the criticism of Iparraguirre's work, commentary about Demitrópulos focuses on the challenge she makes to official history and on her representation of women.

Un piano does feature prominently the story of an immigrant woman, but it also contains a unique representation of the transformation of place. The location is again the southern tip of the Southern Cone, particularly the small ports and harbors where seamen and shopkeepers gather. The cast is a motley crew drawn from the margins of society: a homosexual, a Peruvian bar owner, a frigid Englishwoman bought by an Austrian colonist and sheep farmer, a rough sea captain, Ona and Yámana Indians, a prostitute-nun, and a sailor "gone native" living in a bay with indigenous consorts. Demitrópulos, like Iparraguirre, rewrites the history of Tierra del Fuego "from below." She also articulates a critique of violence with ecofeminist overtones, for she explicitly links the hunger for power over women with lust after new territories. Ultimately, her narrative structure insists that the reader connect this "will to power" in the past to a future laced with the destruction and mourning unleashed by a right-wing dictatorship.

The narrative is a fragmented, at times delirious discourse that follows the obsessions of ship captain Gin-Whisky. During a stay in port, Gin-Whisky becomes infatuated with Nancy, a poor Englishwoman who plays the piano in a seaport bar. Gin-Whisky is also persistently fascinated with his friend "gone native," Isidoro Prutt. Gin-Whisky had sailed with Isidoro Prutt on many voyages until Prutt retired from the sea and set up residence on solitary Desolation Bay. Prutt's abandonment of Gin-Whisky provokes a crisis of personal and collective consequence.

Demitrópulos' novel contests the rhetoric of empire with a complicated narrative that charts the messy imperial projects in Tierra del Fuego. In liberal rhetoric in Argentina and in the metropolitan, colonial/modern rhetoric of nineteenth-century Europe, the march into territories of others was described in grandiose terms, as civilization swept into realms of barbarism. Demitrópulos focuses on the prehistory and aftermath of the advancing capitalist frontier: the poverty and hunger that drove Englishwomen like Nancy abroad and the terrible havoc wrought upon native peoples and landscapes in the march to modernization. Ultimately, Demitrópulos posits that the will to power of empire is a specific, historical, and patriarchal desire, rooted in the ideologies of industrial northern Europe and deployed across immense territories. In moments, the happenstances of empire produce absurdities as great as a piano, grand symbol in the novel of incongruous Victorian bourgeois aesthetics, dragged ashore in a solitary bay miles from England. By means of both metaphor and plot, her story strips imperial logic of its naturalness by juxtaposing ideologies to reveal the disastrous results for people and places.

Demitrópulos consistently shapes her critique of empire with an eye toward gender and place. For example, Nancy is the object of Gin-Whisky's desire, at least in part because the captain is plagued by doubts about his own masculinity. For her part, Nancy is a poor Englishwoman who once worked a factory job in industrialized England.[15] Like many other displaced and disposable workers, she immigrates to South America in search of a better life. More than anything else, Nancy hopes to escape the sexual exploitation of her body and the commodification of her labor that she experienced in the capitalistic, patriarchal order of nineteenth-century England. Instead, she finds herself caught in the same culture of commerce and male privilege across the ocean.

Unfortunately for her, Nancy flees to a Southern Cone nation whose nineteenth-century leaders sought economic and social progress through export-led growth (Bulmer-Thomas 119). European immigrant women were valuable in this context because they labored in agricultural enterprises that formed the basis of this economy and because they had the potential to increase the population, as the (white) mothers of soldiers and laborers. In Argentina in particular, population growth was crucial for economic and political reasons, especially given the loss of life incurred by a century of wars for independence, over national politics, and over disputed borders.

In late nineteenth-century Latin America, even politicians who could agree on little else were in agreement that "gobernar es poblar" [to govern is to populate]. As Latin Americanist Doris Sommer notes, two of Argentina's founding fathers, Juan Bautista Alberdi and Domingo Faustino Sarmiento, "agreed, if on little else, on the need to fill up the desert, to make it disappear" ("For Love" 390 footnote). Furthermore, as Francine Masiello points out, with the expression "gobernar es poblar," "the early liberal intelligentsia demanded a purification of the race and sought a republican metaphor to protect the land from the expansion of indigenous peoples and undesirable 'others'" (5). European women were thus imported and regulated for their reproductive potential. Gender politics played an important role for a state that wanted to convert massive tracts of land, inhabited by native peoples and gauchos, toward uses that were "productive" for an export-oriented economy.

Demitrópulos tells the story of male domination of landscapes and gendered Others in nineteenth-century Latin America through Nancy and the gay bar owner. Both are victimized by their own societies because of sexual difference. Even by running to what she perceived as "the ends of the earth," Nancy could not escape the clutches of the modern world system that worked to bring people and natural resources into a capitalist economy. A working-class woman with little future in England, Nancy was bought as a sexual consort and farmhand by a colonizer with a herd of sheep and a hovel in Tierra del Fuego. Even though an ocean separated her from Victorian England, in Argentina she is exploited and commodified for her productive and reproductive potential. Moreover, she is still part of the same economy of textile production. In England, she worked in a factory; once coupled with the Austrian colonist in Argentina, she tends sheep that provide fibers for export to such factories.

For his part, the Peruvian bar owner was run out of his Andean town by his family. His sexual orientation jeopardized the social order and the economic well-being of the family in a conservative agrarian town. No one would dare marry the sisters of a gay man, and the family would face ruin because economic well-being depended on large, extended family ties.

Nancy is the figure through which Demitrópulos critiques the benefits of empire for its own subjects. Using the lens of gender, Demitrópulos points out that modernity excludes working-class women from many of its benefits, both in Victorian England and in the new Latin American republics. Extractive economies that benefit empires depend upon large numbers of expendable surplus laborers in the metropolis and lands overseas; many of these are women who toil in factories and fields. Demitrópulos gives a close reading of this moment of history, signaling the role of poor, metropolitan women in the large-scale conversion of lands worldwide to the production of fibers instead of foodstuffs.

Detailed, carefully narrated scenes of violence are the instruments by which Demitrópulos uncovers the darker side of the Victorian era. Demitrópulos integrates scenes of violence against animals to force the reader's attention upon the interconnectedness of luxury and status and an economy dependent on brutal exploitation. The lucrative trade in the pelts of the South American fur seal (*Arctocephalus australis*) is Demitrópulos' prime example of how cosmopolitan, nineteenth-century fashions and aesthetics depend upon atrocities unleashed by colonial enterprises.

The passage below, for example, is told in the voice of Gin-Whisky, the opportunistic, licentious sailor. The narrative intentionally juxtaposes a sense of wonder with the rhetoric of commodification. With the mention of money, the narrative abruptly comes into focus, as specific acts of violence follow:

> El espectáculo era fascinante: cientos de lobos de dos pelos se revolcaban al sol, dormitando, y cubrían la extensión de la playa. ¿Cuánta plata había ahí en cueros? Con que cazáramos la mitad ya éramos ricos. A una señal mis hombres se pusieron de pie y a una orden caíamos todos juntos y con el mismo ímpetu desde arriba sobre los durmientes de la playa. Empezamos a golpear en forma seca y sin descanso sobre la cabeza de los animales semidormidos. . . . Así que golpeo, golpeo en medio de aullidos. El eco de los palazos sobre las cabezas suena seco y opaco (81).
>
> [The spectacle was fascinating: hundreds of fur seals rolled around in the sun, dozing, and they covered the length of the beach. How much money was there in pelts? If we only got half of them, we'd be rich. At a

signal, my men stood up and when the order sounded, we all fell upon the sleeping ones with the same impulse. We started beating the heads of the sleeping animals suddenly, ceaselessly. . . . So, I beat, beat, in the midst of howls. The echo of the blows on their heads sounds hollow and opaque.]

The opening sentence of the passage communicates awe, but the rest of the passage disavows any notion of romantic wonder before the grandeur of nature. The focalization of the passage becomes clear in the second and third sentences. Gin-Whisky does not see seals but money. The next sentences detail the violence against animals; the use of the first-person plural verbs reference the group of men, but they also uncomfortably read as if to include the reader.

Demitrópulos also emphasizes the dehumanizing effect of the violence on the men who carry it out. Just as they disregard the howls of the animals they are beating to death, they also ignore the scream of their comrade as fleeing seals sweep him into the water. They stop only when there are no more seals left alive, and their cohort is long dead then.

Demitrópulos uses vivid language to force the reader to accompany the men in their acts. Verbs emphasize tactile experiences ("golpeo, golpeo"), sound ("el eco de los palizos"), and sight. Preceding the passage of violence in the narrative are pages detailing the men's planning and a paragraph detailing the seals' organization, habits, and environment. The build-up to the violence is inexorable.

With this, Demitrópulos portrays the seal hunters as the first line of assault in an imperial enterprise. Their violence is part of a larger war to bring the unique ecological and cultural world of the austral islands into the modern world system. Who exactly are the seal hunters, though? They are freelance agents of violence and change, and yet Demitrópulos also makes it clear that they are the disposable sons of the metropolis. In Europe, there is no work for them that delivers the promises of capitalism; there is little prosperity and upward mobility for them in Amsterdam, London, or Liverpool. The colonial enterprise in the form of commercial fleets is a safety valve for the surplus labor of industrial Europe. The *loberos* are the vehicle through which Demitrópulos critiques the false consciousness of capitalism. The men have bought into the narrative of progress and wealth; they are willing participants in their own exploitation and dehumanization. They undertake risky voyages to club seals and transport contraband because they are eager to share in the symbolic currency of wealth: abundant liquor, unregulated sex, freedom of movement. However, the narrative makes clear that their profits disappear into the pockets of bar owners, pimps, and the creditors who finance their expeditions. They have not escaped misery, Demitrópulos suggests, because misery is inescapable once the men accept the lure of sharing in easy fortune, the false promise of global capitalism.

Demitrópulos ends the novel in a way that links the nineteenth-century colonial enterprise and the birth of global capitalism with late twentieth-century, right-wing dictatorship and the launch of neoliberal initiatives in Latin America. In the final chapters of the book, her characters, like those of Iparraguirre, must all come to terms with the global reach of the new world order, whether they embraced it or not. At the end of Part II, the broke and broken ship captain, Gin-Whisky, points his ship toward Bahía Desolación to seek out his friend gone native, Isidoro. His declared purpose is to deliver to Isidoro the piano Nancy left behind when she returned to England, but his unspoken intentions appear to be tied to a repressed attraction to his friend.

Isidoro is the narrator of all but the final pages of Part III, the last segment of the novel. For Isidoro, the unsolicited delivery of the piano and the arrival of Gin-Whisky are unwelcome intrusions in the community where he has found some peace, outside of history, as he puts it:

> Me sentí perdido y traté de explicarle otra vez: oye, Gin-Whisky, aquí el pasado no existe, tampoco el futuro, solo vivimos el presente; me he asimilado a estos indios y estoy bien así, esto es lo que estuve buscando (155).
> [I felt lost, and I tried to explain it to him again: look, Gin-Whisky, the past doesn't exist here, neither does the future; we only live in the present. I've become assimilated to these Indians and I'm fine here, this is what I was looking for.]

Here Demitrópulos sets up a final critical conflict between two participants in the modernizing enterprise. Isidoro has broken free after a life at sea serving Gin-Whisky. Gin-Whisky finds the decision baffling and deeply disturbing. Isidoro's life of subsistence challenges the core of Gin-Whisky's worldview, and its very existence is a menace to his sense of self.

In Bahía Desolación, a spontaneous community has formed out of the marginalized rejects of two societies. Isidoro welcomes and cohabitates with Fuegian women who have escaped from the violence of seamen and their own people. Among them, Isidoro enjoys some patriarchal privilege (the women care for the children and attend to the food), but it is interesting that Demitrópulos depicts the community as the only site of consensual, noncommercial sexual activity in the novel. Isidoro's partners come to the bay of their own will, and they organize their lives and those of their children with Isidoro according to their customs. Sexual politics in the bay are different, as are the politics of economics and leisure. Isidoro has fully retreated from modernity. He spends his time living with his wives instead of finding solace in the company of seamen, prostitutes, or pianos in seaport bars. This transgression is incomprehensible to Gin-Whisky, who plays his role in the global, capitalist enterprise unquestioningly until the end.

In her conclusion, Demitrópulos crafts a final confrontation between modernity and the alternative world of Isidoro. This denouement progresses with a relentless violence that ends in madness and an uncertain future. The arrival of Gin-Whisky first brings doubts about the past and dismay as to the future; it jolts Isidoro out of the present before it disrupts the social order. Gin-Whisky appropriates all of Isidoro's indigenous companions as his own sexual consorts, and Isidoro finds himself on the margins of a community he created collaboratively with the native women. Finally, the confrontation brings death in the form of the piano.

Unloaded from the ship at Gin-Whisky's insistence, the piano is an unwelcome intruder. Consider the following passage, narrated by Isidoro, in which Demitrópulos contrasts Gin-Whisky's romantic notion of the musical arts with Isidoro's appreciation for the natural music of Tierra del Fuego:

¿Música?, respondo. El mar, las hojas, los pájaros, tienen música. El agua que cae de la cascada, la conversación de mis mujeres, produce música. No necesito ninguna otra fuera de los alciones y petreles, o de las gaviotas volando por el cielo (165).
[Music? I respond. The sea, leaves, birds, make music. Water that tumbles from the falls, the women's conversation, these produce music. I don't need anything other than the kingfishers and gulls flying in the sky.]

Ultimately, the piano is more than an intruder; it tolls the death knell of the community. The piano takes up space, disrupts the rhythm of life, and draws in a sick ship captain and his crew, who hear the noise and dock. Their contagion, a debilitating fever, spreads long after they are gone and devastates the community.

Using the piano as a leitmotif throughout the text, Demitrópulos has seized on the perfect symbol of Victorian bourgeois culture. Nancy learned a fondness for the piano in the parlor of her English home, and later her melodies invited Gin-Whisky's attentions in the bar. The last section of the novel emphasizes the dissonance of the piano in Bahía Desolación. Shortly after the arrival of the ship, Gin-Whisky moves on in search of his own fortune, taking over the mission from the sick captain to deliver settlers (most of whom are retired sailors) to the Malvinas. Eventually, Gin-Whisky fails at reading the icy landscape and steers his ship in circles. With passengers and crew delirious from hunger, it sinks in the ice floe, lost en route to the Malvinas.

For readers familiar with Argentine history, this conclusion suggests the catastrophic conflict between the military dictatorship and the British over sovereignty in Malvinas in the 1980s. Blinded by ambition and eager to hold onto power like Gin-Whisky, junta leaders provoked a conflict that sent hundreds of sailors to their deaths and cost the country a fortune. Ultimately, it also brought an end to the right-wing dictatorship.

In the final pages of the novel, Demitrópulos employs shifting focalization and jarring ruptures to create the sense that the characters' lives are spiraling out of control. Human beings have run into the physical limits of the natural order: illness and ice. While Gin-Whisky sails off into the ice floe, the contagion brought by the intruders quickly ends the lives of all but two members of the community (Isidoro and a mestizo son). Ultimately, Isidoro also falls ill, attended only by a young son, and the destruction of the bay community appears complete.

Isidoro's fever-induced delirium is the textual device by which Demitrópulos connects nineteenth-century colonial violence with twentieth-century dictatorship. The last pages of the novel remind readers of a recent chapter in the history of Argentina. In the early 1980s, during the dictatorship, divisions within the military, denunciations by human rights groups, and economic woes plagued the right-wing military officials in power. In an effort to distract from these issues, the military government provoked a dispute with Great Britain over claims in the Malvinas, or Falkland Islands. Events rapidly escalated in 1982 to outright war. The ultimate outcome of this war was the collapse of the military junta, at the cost of hundreds of lives and millions of dollars.

In the last pages of Demitrópulos' novel, after reflecting on his own "barbarism," delirious Isidoro describes Buenos Aires like this:

> Buenos Aires. Polis de cambalache. First is money, decían en la calle Florida. Pero ¿no era que time is money? ¿Quién entiende a esta república republicana? . . . Corro desnudo por calles desiertas de una ciudad extraña. ¿O por las de Buenos Aires sucias de sangre? Llego a una plaza y una procesión de mujeres me sorprende. Llego a una plaza y por una diagonal, esgrimiendo signos e imprecando, sale la confusa procesión de mujeres. . . . Por otra diagonal avanzan hombres con máscaras del carnaval de Oruro. Danzan entre enanos, demonios y deseos prohibidos. ¿La ficción civilizada se ha vuelto barbarie? (199–200)
> [Buenos Aires. City of trade. First is money, they say on Florida Street. But, wasn't it that time is money? Who understands this republican republic? . . . I run naked through the deserted streets of a strange city. Or those of Buenos Aires covered in blood? I come to a plaza and a procession of women catches me by surprise. I come to a plaza and across the diagonal, wielding signs and cursing, comes a confused procession of women. . . . Along the other diagonal men advance wearing the masks of the carnival of Oruro. They dance among dwarves, demons, and forbidden desires. Has the fiction of civilization turned into barbarism?]

The passage (abbreviated above) begins with a reference to the capitalist realities of modern Buenos Aires. It then turns to an image of violence, of blood running through the streets. Finally, the reference to a procession of women in a plaza, a nod to the Madres de la Plaza de Mayo, directs the reader's attention to Isidoro's future and Argentina's recent past of dictatorship, the Dirty War, and neoliberal reform.[16]

Demitrópulos here directs the reader's attention to a pressing question about the nature of civilization and barbarism: Did the seeds of civilization sown in Gin-Whisky's era grow into the brutality of dictatorship?

This conclusion bears further analysis from an ecocritical perspective.

Isidoro hallucinates about the capital, the heart of the modernizing enterprise in Argentina and point of pride to this day, touted for feeling like a "European city."

At what cost, though, did Buenos Aires flourish? The Demitrópulos novel insists on contextualizing the flourishing, modern capital within a legacy of violence. Buenos Aires grew tremendously in the last decades of the nineteenth century; in fact, an American researcher noted in 1900 that Buenos Aires was the largest Spanish-speaking city in the world, and he marveled that "more than half of its citizens are foreign born and the city itself is fast losing much of its Spanish character" (Carpenter 309–10). A trans-Atlantic cable linked Buenos Aires to London in 1874, and the city thrived as a center of trade and port for the influx of immigrants (Winseck and Pike 71). From this "polis de cambalache" emanated cultural and environmental transformations of landscapes well beyond the confines of the capital.

Sailors, immigrants, and merchants drawn to the busy port introduced new tastes in the furthest reaches of the continent. So, too, did global communications networks: "Speedy connections to the metropolitan centers of the world—London, Paris, New York—meant that such places now loomed larger in the imaginations of local and transnational elites who looked to them for their cues on style, politics, the design of urban spaces and architecture, and their ideals of modernization and progress" (Winseck and Pike 71). Demitrópulos references the predilection for European luxury items in Gin-Whisky's very name; his consumption defines his identity as a seaman. In reality, government policies, financial institutions, and advertising all shaped consumer preferences in territories suddenly brought into the modernizing project.

Demitrópulos ends her novel grimly with few survivors. The dark conclusion brings concerns about modernization in the 1800s into conversation with the tragic consequences of dictatorship in the twentieth century. As in Iparraguirre's novel, two marginalized characters still stand in the last pages of the text. For Iparraguirre, Guevara and Graciana survive to tell the tale. Writing less than a decade after the fall of the dictatorship, Demitrópulos does not offer readers a figure like Guevara to sort through history. Instead, Isidoro destroys his diary full of memories and lies about a troubled past.

Emerging from sickness into a Fuegian spring, Isidoro first observes his son John Yámana, whom he now sees as an individual: "muchacho de talones fuertes, futuro caminador, brote del paisaje que retarda lo increado en latencia, alguien que puede destruir la ficción que a mí me ha destruido, el único que sabe esperar la integridad" (202) [a boy with strong heels, a future walker, sprung from the landscape that retards what is not yet created, someone who can destroy the fiction that has destroyed me, the only one who knows how to wait for integrity]. Next, Isidoro destroys the diary and burns the pencil with which he wrote it. Finally, he hacks away at the piano and invites his son to join him: "Buena leña—dijo John Yámana. Y encendió una fogata" (203) ["Good firewood," remarked John. And he lit a fire].

This comment by John is the last glimpse Demitrópulos gives of the bay community. John and Isidoro disregard its artifice and its symbolic value, and the piano returns to its natural state. The wood serves one of its primal functions for humans: warmth and heat. Isidoro and John thus rid the bay of the symbol of the plague, and the fire signals the renewal of life for them.

In Demitrópulos' conclusion, Isidoro must destroy his connection with the past and forge a means of survival with his son. Gin-Whisky dies, taking the ill-fated settlers down with him. Demitrópulos' novel thus forcefully draws the reader from the critique of colonialism and emergent globalization in the Americas to a denunciation of both state terror and unrestrained capitalism under dictatorship. Her representation suggests that what began in efforts to exploit land, women, and native peoples ends in modern Argentina, with a slaughter that makes blood run through the capital. An enforced patriarchal culture, exploitation of women, brutality toward nonhuman nature, dehumanization of the brutalizers—all go hand in hand in Demitrópulos' modernity.

Fuegia

Like Iparraguirre and Demitrópulos, Belgrano Rawson uses the history of Tierra del Fuego to critique modernization and draw attention to its toll on cultural and biological diversity. Belgrano Rawson focuses particularly on the violent endgame of Argentine governmental and British imperial projects at the turn of the twentieth century. By this time, both Britain and Argentina had pushed the capitalist frontier well into Tierra del Fuego. Driven by policies pushing export-led growth, the Argentine government was eager to populate Patagonia and beyond with European immigrants willing to homestead there. As a result, British landowners gained extensive land concessions from the Argentine government and enclosed territories for the booming sheep industry.

The effect of colonization on the Fuegian people was disastrous, as anthropologist Anne Chapman notes:

> The Selk'nam (Ona) people, inhabitants of the largest island of Tierra del Fuego in the extreme south of South America, were still living in accordance with their ancient traditions when, in 1880, the White man suddenly began to colonize their land. During the final decades of the nineteenth century an unknown number of Selk'nam were massacred by the newcomers. Then and later, countless others died of imported diseases, against which they had no natural immunity.... In 1980, one hundred years after the beginning of colonization, only one or two direct descendents of these people remain (1).

Belgrano Rawson recovers the memory of the indigenous peoples of the canals and islands of Tierra del Fuego during a moment of intense competition for resources and livelihood. He recuperates this history in a drama about one fictional indigenous family displaced from their "culture of habitat" by modernizing processes. At the end of his story, the family no longer exists as an integral social unit, but the cycle of violence catalyzed by colonization persists.

Literary critics, particularly Argentine commentators, have paid attention to Belgrano Rawson's novel because of its representation of violence and cultural loss. For example, Nora Longhini, Márgara Averbach, and Norma Pérez Martin interpret the novel in the context of memory and the erasure of the indigenous past in Argentine national history. Their criticism is grounded in a broader field of inquiry into memory in post-dictatorial fiction. Though her focus is generally on marginalization, Corina Mathieu also mentions the connection between environmental degradation and cultural loss: "La llegada de los blancos dispuestos a explotar sin tregua los recursos naturales determina que los nativos debieran enfrentar, con el transcurrir del tiempo, no sólo la natural inclemencia del clima austral, sino también la escasez de alimentos" (149) [The arrival of whites predisposed to exploit natural resources without halt, meant that over time, natives would face not just the natural inclemency of austral climate, but also a shortage of food]. To date, though, no full-length ecocritical analysis of Belgrano Rawson's novel has been published.

Critics have hailed his novel for its recovery of memory, but Belgrano Rawson has made repeated protestations in interviews that *Fuegia* is not a historical novel ("Escribir de oído"). Nevertheless, as Longhini notes, historical, anthropological, and ecological details of the plot are solid and based on exhaustive archival research (77). The author himself describes a process of composition in which he immersed himself in documents about the region and then distanced himself from them in order to create a story that would be true to the spirit of historical events. He also accompanied a team of biologists on a scientific trek across the region.

One way Belgrano Rawson communicates place and the spirit of events is through fictionalized place names. The novel never once uses the term "Tierra del Fuego," and nearly all the places described in *Fuegia* have fictional names. The representation of the cultural and ecological transformations is accurate, but the author has intentionally left the identities of places and people unmarked by modern nomenclature.[17]

The ambiguity about place names provides a valuable opportunity to open an ecocritical conversation about the novel. Some critics, like Mathieu, have asserted that the erasure of real place names makes for a more universal story: "al crear una ficción cuyos entornos, aunque reconocibles, no encajan exactamente dentro de ningún marco real, ésta asume un carácter simbólico que le confiere mayor universalidad" (147) [by creating a fiction whose geography is recognizable but does not exactly fit within any real parameters, this takes on a symbolic function that confers more universality to the story]. The effect of the erasure goes beyond that, though; it accomplishes the removal of Western cultural markers for places. European explorers and cartographers gave Tierra del Fuego the names it bears on maps today (the "Beagle Channel" and "Cerro Darwin" [Darwin's Peak], for example). With the extinction of other cultural groups, native place names were lost. By replacing European nomenclature with new referential terms, like "el país de las lluvias perpetuas" [land of perpetual rain], Belgrano Rawson draws attention to a different way of perceiving landscape. Readers must reimagine the landscape according to its natural rhythms (felt by longtime inhabitants) rather than by the egocentric nomenclature Europeans gave to it on their maps of the territory.

Place in *Fuegia* is central to the plot from the first page of the novel, which describes the brief return in winter of hungry guanacos (*Lama guanicoe*) to coastal grazing lands. Even here, Belgrano Rawson hints at two types of violence: that of the hunt by indigenous peoples and that violence lurking offshore in commercial vessels. In descriptions of the landscape and narrative about the actions of those who change it, Belgrano Rawson points to the clash of cultures as modernity transformed Latin America.

The modernizing project took shape in Argentina through a collaboration by creole elites with British investors, the vanguard of what Jennifer French calls the "Invisible Empire" (7–8). With naval supremacy and vast commercial ambitions, in the late nineteenth and early twentieth centuries, the British exerted influence in Argentina through financial institutions, commercial storage and shipping, and concessions for railroad construction. Along with British capital, new people, goods, and patterns of consumption all came to the Americas. This British investment was not viewed suspiciously by the Argentine elite until much later.

Indeed Argentine national policy promoted foreign investment, as well as immigration to Argentina from continental Europe and the British Isles. Immigration efforts were so successful that by 1914, the foreign-born residents constituted one-third of the population of Buenos Aires and three-fifths of Santa Fe (Rock 167). In economic policy, the Argentine government pursued export-led growth through expansion of production of agricultural commodities. Efficient commodity export necessitated huge sums of capital investment in infrastructure projects like railroads, ports, and communication. Taken together, British and Argentine policies converged to push new people, crops, and livestock into territories that had, until then, provided habitat and sanctuary for indigenous peoples. In Argentina, the frontier of capitalism and liberal politics moved inland and southward, and clashes occurred when the peoples already inhabiting these lands responded to the encroachment.

Belgrano Rawson depicts turn-of-the-century events to draw attention to the violent details of national consolidation and capitalist expansion in Latin America. Debates about indigenous affairs in fact raged in Buenos Aires at the time, and the Argentine Congress deliberated over whether the indigenous people were citizens of the Republic and thus worthy of protection. Indeed, earlier Argentine thinkers like Domingo Faustino Sarmiento had preferred truces and educational campaigns as ways of settling tensions with indigenous peoples. By the early twentieth century, there was considerable diversity of thought in Argentina about native peoples. Landowners on the frontier, however, complained persistently about profit losses stemming from Indian raids on herds of livestock. At least part of the debate in Buenos Aires about Tierra del Fuego was concerned with escalating violence between British settlers and native populations (Chapman 11).

Belgrano Rawson's intimate representation of people and landscapes draws attention to the "double erasure" of indigenous peoples in Argentina. Native peoples were not just physically eliminated from the national scene, they were also expunged from collective memory. Susana Rotker observes that "to the exterminated Indians not even the myth of origins was conceded, and it is a rare Argentine history that begins much before the period of independence from Spain" (21). Magdalena Perkowska-Alvarez has asserted that *Fuegia* is

> una version fueguina de la Conquista del Desierto que también inscribe una doble borradura: la aniquilación física de los pueblos encarnados por los personajes indígenas de la novela y la eliminación de estos sucesos de la memoria histórica por parte de los responsables o las autoridades (argentinas y chilenas) que se beneficiaban del progreso implementado por los colonos en las islas (85).
>
> [a Fuegian version of the "Conquest of the Desert" that inscribes a double erasure: the physical annihilation of the communities comprised of the indigenous characters in the novel and the elimination of these events from historical memory by those responsible and by those authorities (Argentine and Chilean) who benefitted from the progress colonists brought to the isles.]

In *Fuegia*, Belgrano Rawson reminds readers of the tremendous cost of an Argentine national image that is more European than Latin American. The filling of "empty" spaces, as in the Conquest of the Desert or colonization of Tierra del Fuego, meant the elimination of native grasslands, as well as alternative modes of human existence. Belgrano Rawson peers behind the liberal discourse of the nineteenth century into so-called "empty spaces" and guides readers to see a landscape that rhetoric metaphorically eliminated.

According to nation-builders like Sarmiento, blank spaces were territories with enormous economic potential. In actuality, of course, blank spaces are ecosystems, oftentimes sparsely inhabited by both sedentary and nomadic indigenous cultures, neither of which generally exceeds the carrying capacity of the land. Archival documents and photos from the time period suggest daily life along the Argentine frontier was a complicated reality of inter-cultural collaboration and conflict. Nevertheless, as domestic populations and consumption increased in Argentina, the attendant cultural and envi-ronmental changes were momentous and enduring. The national image of a homeland that was not Latin American but European came with a heavy ecological and cultural price.

Belgrano Rawson's novel features violence against people, animals, lands, and cultures to explicitly connect the pursuit of profits and the use of land to the elimination of cultures. Eleven nonlinear chapters outline a plot about the parties involved in a massacre of indigenous people at Lackawana. Through-out, all individuals are humanized, even those that grow into rapists, thieves, or murderers. The novel does justice to the complexity of human beings, but it does not make excuses or justify atrocities. Rather, it repeatedly under-scores the cultural and social conditions that lead to violence and exploita-tion. The novel also relates the memory (on the part of each character) of an earlier state of well-being, usually during childhood, before the individuals were thrust into their roles in modernity.[18]

By depicting three generations of a Fuegian family, Belgrano Rawson af-fords readers a view of environmental and cultural change across a contin-uum of history. The family in focus is that of Camilena (a *canoera*, Belgrano Rawson's term for the Yámana) and Tatesh (a *parrikens*, or Ona/Selk'nam), and their three children. In various flashbacks, Camilena and Tatesh also re-member their own parents, and this textual device allows Belgrano Rawson to probe an earlier period of colonization.

Though the focus is on one indigenous family, the cast of characters of the novel is broad: a Chilean doctor and his daughter; Mrs. Dobson, the elderly Englishwoman who lives at the mission her late husband founded; Father Lorenzo, a Catholic priest with genuine concern for the indigenous people; Thomas Jeremy Larch, a British sheep farmer; tourists rounding the Cape; and a couple of shipwrecked sailors. For the indigenous protagonists, the lives and ambitions of these outsiders converge with drastic consequences. Over the span of three or four generations, Fuegians found their lives threatened by the encroachment of seal hunters and whalers, the growth of sheep farming, and ultimately disease, forced conscription and relocation, and out-right violence.

Belgrano Rawson focuses on the last act of this tragedy in his novel. The first chapter ends with a bloody crime scene: his neighbors find the elderly Thomas Larch dead in a pool of blood, and his servant Beltrán or Lucca, Camilena's youngest son, has disappeared. In a dizzying array of analepsis and prolepsis, Belgrano Rawson ties the murder of Larch to the acts of vio-lence that precede it, such as sexual abuse of indigenous women at the Prot-estant mission, cultural plunder by naturalists, the rape of Camilena, and the massacre at Lackawana ordered by Larch, the sole survivor of which is Lucca (renamed Beltrán and given to Larch as a servant).

Rape and murder are the bookends of violence in the story Belgrano Raw-son tells. One of the earliest actions of the diegesis is the rape of Camilena by sailors from a small commercial vessel, an attack that unleashes a chain of events leading to their massacre. A summary of these events, in the order in which they occur, will be helpful for the analysis to follow.

In the chaotic moments before the assault, Camilena's small children flee into the woods, and her husband presumably later finds them. The family ul-timately reunites at Abingdon, the missionary enclave where Camilena spent her childhood. Faced with a measles outbreak at the mission, Camilena and Tatesh seek a document from Mrs. Dobson, the widowed missionary, to guarantee safe passage north through enclosed lands. Tatesh plans to take his family to his own people who live further north, retracing the journey he made with his desperate mother more than a decade before. Mrs. Dob-son refuses him such a document, but the family leaves anyway, fearing the epidemic. On their journey northward, the family stumbles across the ship-wrecked sailors who raped Camilena, and they attack them in revenge. Fi-nally, Camilena and her family themselves die at the hands of Larch and his men, who hunt them down with vicious dogs because the family killed sheep in the enclosed land through which they passed. Only Lucca, the youngest son, survives, and the man responsible for the massacre at Lackawana keeps him as a servant. Neither the journey of Tatesh's mother nor that of Tatesh himself would be completed. Tatesh's mother ends up a servant in Buenos Aires, and Tatesh dies with his family in Lackawana.

Belgrano Rawson presents the arrival of modernity in Tierra del Fuego as a convergence of racism, sexism, and unrestrained capitalism. The Fuegian landscape and its subjugation to these forces is integral in the narrative from the first page. The opening chapter of *Fuegia* reads like a lyrical environmental history of Tierra del Fuego after European contact. Indeed, the native fauna is the first image introduced. A herd of cautious guanacos approaches the coast, wary of attacks from the indigenous people who have themselves already disappeared. The guanacos are also greatly diminished, and Belgrano Rawson personifies them as animals with a memory of their own conflicts and battles. The chapter proceeds to resignify the Fuegian landscape place by place, marking each with an indication of the dramatic change it experienced. For instance, the coves and bays once served as refuges for seal hunters, but "ya casi no había lobos y los loberos andaban en la última miseria" (19) [there were hardly any seals anymore, and the seal hunters lived in abject poverty]. As in other ecologically imaginative novels from Latin America, Belgrano Rawson's novel emphasizes the interdependent fates of humans and nonhuman nature.

Belgrano Rawson weaves together environmental and social history to narrate the tendency of modernity to eliminate both cultural and biological diversity. When seals were plentiful, Belgrano Rawson observes, the *loberos* ignored native inhabitants and spent their time hunting seals on the rocky Atlantic outcroppings. Later, though, desperate *loberos* turned to less lucrative endeavors for survival. They ventured further into Fuegian territory, hunting penguins for oil, robbing indigenous peoples for food, and raping women for diversion and a sense of control.

At the same time that attacks on indigenous life came from sea, the capitalist frontier moved in by land. To tell this story, Belgrano Rawson offers one emblematic example, a memory focalized, or filtered, through Tatesh. Tatesh vividly remembers his father's shame at being earless. Rather than explain the violence behind this condition directly, Belgrano Rawson disperses details throughout the text to reveal that the removal of the ears was an act committed by mercenaries paid to hunt Fuegians. Historical documents confirm the practice, in which hunters of humans were paid one pound sterling for proof of the demise of their prey (in this case, something portable and lightweight: the ears).

Another narrative strategy Belgrano Rawson employs to unsettle traditional conventions is marking the passage of years by charting changes to the ecosystem. Rather than using Western markers of time, he highlights instead markers of environmental degradations and its cultural consequences for native peoples. One passage, focalized through Tatesh, notes that "Era la época en que su padre aún cazaba guanacos y no debían andar mendigando" (139) [It was when his father still hunted guanacos and didn't have to go around begging]. In statements like these, Belgrano Rawson explicitly links the exhaustion of natural resources to the creation of new social subjects. As the territory sustains a larger population of sheep valuable for profit, grazing lands for native species like the guanaco diminish. With resources depleted and lands enclosed, indigenous peoples must turn to wage employment or banditry to survive. Their only option is to work for those who have displaced them from their livelihoods or to supplement an existence on the diminishing margins of land by stealing sheep for food. Ultimately, there is no room for people like Tatesh to survive as integral familial or cultural units.

The progression toward the violent destruction of the family is constant, but readers may only fully comprehend it by sorting though a proliferation of details and memories offered in nonlinear narrative. For example, readers encounter the last chapter of Tatesh and Camilena's story as one of the first events of the narrative. It is the revenge of Lucca, the toddler who survived the massacre at Lackawana. Kept for years as a servant and asked by Larch to show off his father's skull for tourists, one day Lucca kills Larch. Neighbors find him in a pool of blood with his testicles in his mouth, and his dog lying dead by his side.

Larch's emasculation is foreshadowed in the text; at one point Tatesh's father had been employed to castrate sheep for the landowners. The encroachment of sheep eventually cost the entire family their lives, and young Lucca's first awareness of the impending massacre at Lackawana was the sight of his beloved dog dead by his side, killed by the curs that, goaded by Larch's men, next attacked his siblings, parents, and others. Lucca's violent revenge points toward the patriarchal privilege that drove arrogant interlopers to expand personal empires no matter the cost to others. After the murder, Lucca simply disappears, another instance of the retreat from modernity in yet another novel. In this case, it is also a gesture toward the disappearance of Fuegian people from contemporary life.

Throughout the novel, Belgrano Rawson links modern landscapes and cultures to a history that created them. For example, Belgrano Rawson shows how penguin oil comes from the mass extermination of these animals and the boiling of their bodies in cauldrons. Pasture land comes from the physical elimination of indigenous adults and the children they have hidden according to custom before the final battle at Lackawana. *Fuegia* presents the history of modernity in Tierra del Fuego as a series of episodic acts of violence against the bodies of humans and nonhuman animals, and in so doing, creates a tone of dread and foreboding. Readers know by constant references to the event that a massacre at Lackawana will occur and that it will conclude with an orphaned Lucca.

Details of the suffering of Tatesh and Camilena invite readers to identify with them and to will away the violent approximation of modernity. Here, too, violence against animals has a narratological function. Averbach asserts that in *Fuegia,* "la acción tiene su clímax en una masacre de seres humanos (muchos de los cuales son niños) y las muertes anteriores de guanacos y pingüinos funcionan en parte como anticipatorios de un espanto todavía mayor" (66) [the action reaches its climax in a massacre of human beings (many of whom are children) and the previous deaths of guanacos and penguins serves in part to foreshadow an even greater shock]. The arrival of "civilization" and modernity in Tierra del Fuego is messy and brutal. Vivid details of violence illuminate the quotidian realities largely erased from historical memory.

Belgrano Rawson uses narrative devices that require the reader to process and reprocess the story constantly. He offers no figure like Iparraguirre's Guevara, and his conclusions are even more dismal than those of Demitrópulos, in which the mestizo son and father survive to face an ambiguous future. The last pages of *Fuegia* provide no clue about the future; like Demitrópulos' story, they feature the delirium of one surviving shipwrecked sailor, transported by a ship captain back to Europe and hallucinating about Sunday dinner with his family. It is striking that both *Fuegia* and *Un piano* conclude with scenes of madness and delirium. Ultimately, agents of social and environmental change are caught in violence that exacts a heavy personal toll. They are alienated from other humans and nonhuman nature through acts of violence, the purpose of which is to further the reaches of empire. In both novels, madness is the product of the trauma of modernity.

Conclusions

The ecological imaginations of Iparraguirre, Demitrópulos, and Belgrano Rawson bring cultural and environmental change into the foreground of the narrative about modernity in Latin America. Their choice to set their stories in a territory at the farthest reaches of the British empire and the Argentine nation reveals the extent of the dominance of the global economic system. In a new moment of ecological and social peril that accompanies economic transformation, the authors insist that readers understand the history of global enterprises on the people and places of the Americas. As Iparraguirre's narrator said of Button: "Por alguna razón que desconozco, mi historia no puede explicarse sin la suya." (52) [For some reason I can't fathom, my life's story cannot be explained without his (29).] Neither can any of ours.

Chapter 2

Contests for the Amazon

Perhaps more than any other region of the world, the Amazon basin is synonymous with ecological wonder and environmental crisis. Representations of the Amazon abound in scientific studies, popular media, literature, film, travel writing, and oral histories. These depictions come from all corners of the globe and from people with diverse backgrounds. Some portrayals of Amazonian flora, fauna, and cultures are culturally and scientifically accurate; others more vividly depict the realities of Western imaginations.

This chapter moves from conflicts in turn-of-the-century Tierra del Fuego to contests for the Amazon in the twentieth century. The Amazon is at the center of many controversies: development versus conservation, the future of biofuels, the viability of sustainable harvesting, petroleum exploration, and debt-for-nature deals. The literary works of this chapter fictionalize struggles over the future of the Amazon during distinct historical moments. *Mad Maria* (1980), by Manaus-based author Márcio Souza, considers an era of railroad construction in Brazil during the rubber boom. *Fordlandia* (1997), by Argentine writer Eduardo Sguiglia, examines the business venture of Henry Ford in the Amazon. Finally, *Un viejo que leía novelas de amor* (1989), by Chilean-born author Luis Sepúlveda, is set in the midst of colonization schemes and the emerging petroleum economy in eastern Ecuador.[1]

Each author sets his novel in a moment of accelerating environmental degradation, from early twentieth-century railroad construction (Souza) and rubber extraction (Sguiglia) to an era of petroleum exploration (Sepúlveda).

The ecological imagination of each author is also quite distinct, but they share a common ideological bent. Each author appropriates, replicates, or responds to tropes and fantasies about the Amazon in order to critique the neocolonialist plans of U.S. investors and Latin American elites. The Amazon, and the discourse of nature that evokes it, serves as a vehicle by which to mock and condemn public and private efforts to circumscribe the Amazon within a capitalist empire the authors present as an incongruous imposition.

For this reason, the novels may be problematic for some ecocritics, especially those who follow Glen A. Love in "Revaluing Nature" and eschew literature in which anthropocentric dramas take center stage.[2] Even in *Un viejo que leía novelas de amor*, the most overtly ecocentric narrative, human dramas figure importantly in the retelling of a classic "man versus nature" plot. Anthropocentric stories about the Amazon have their ecocritical merit, though. They illuminate political, racial, and economic entanglements that alter the nonhuman realm of the Amazon. They also represent creative challenges to economic forces that value the region for profit, not for either people or place.

Contextualizing the Novels

This chapter considers three books about the Amazon region, but only one is by an author based in that region. Márcio Souza was born in Manaus in the Brazilian Amazon, and he explores the modern history of the Amazon in writings he still produces in his hometown, a city of millions he temporarily left during the dictatorship. By contrast, neither Sguiglia nor Sepúlveda lives in the Amazon, and Sguiglia's inspiration for writing came from photographs.

Why not include more writers from the Amazon then? The facts behind the reasons are complex, but the answer is simply that there are few published novelists from the Amazon at all. To understand why this is the case, it is necessary to understand political and literary history. Major economic centers in South America have historically been located on the coast or in highland valleys of the Andes. Colonial cities such as Quito, Ecuador; Lima and Cuzco, Peru; São Paulo and Rio de Janeiro, Brazil, still thrive today.

Centers of national and regional power, with important political and cultural scenes, they are all located hundreds of miles from the Amazon. Aspiring writers from the Amazon must overcome huge sociopolitical obstacles, including isolation and prejudices in the arts scene, before they see their works in print. An *International Herald Tribune* piece noted that "until recently [Amazonian writers] were cut off from the main intellectual centers in their own countries, which in turn thought of the Amazon, on those rare occasions when it came to mind, in stereotypical terms" (Rohter). Márcio Souza is one of just two prominent Amazon-born authors from Brazil. The other is Milton Hatoum, but he has resided in São Paulo since he was fifteen years old.

Eduardo Sguiglia is from Argentina, not the Amazon. His inspiration for *Fordlandia* was a small collection of photographs of Fordlandia published in a book (Camara). Sguiglia has served in official capacities as an economist in Argentina, and *Fordlandia* was his first novel.[3] It was well received in Argentina, and when it appeared in English translation, the *Washington Post* book critic identified it as one of the top novels of the year (Yardley).

Luis Sepúlveda has also gained renown for his literature. In 1988, *Un viejo* won the Premio Tigre Juan for best first novel published in Spanish. Latin American critic Ilan Stavans called Sepúlveda's first novel his best because "it finds joy in the art of building literary creatures" and "is proud of its sense of place" (369). Long identified with environmental consciousness, Sepúlvda has an acute ecological imagination he expresses in short stories, essays, and novels. Critic Silvia Casini has gone so far as to state that all of Sepúlveda's novels "ponen el foco en el avasallamiento del planeta por parte de un capitalismo que sólo piensa en beneficios económicos" (104) [put the focus on the subjugation of the planet by a capitalism that only thinks in terms of economic benefits].

Sepúlveda's personal biography reads like the story of the Latin American left over the last four decades. As a young man, Sepúlveda was active in the arts in Chile until the regime of General Augusto Pinochet jailed him in the 1970s. After Amnesty International successfully pressed for his release, Sepúlveda traveled throughout the Americas (much of which was gradually falling or fallen to dictatorship or engaged in civil war) and Europe.[4]

Souza, Sguiglia, and Sepúlveda depict economic, historical, and cultural realities of the Amazon. Each author incorporates varied cultures and peoples of Amazonia, from indigenous residents and internal migrants to American entrepreneurs and trophy-hunting tourists. Though undeniably anthropocentric, the representation of a populated Amazon is a significant contribution to literature. As Candace Slater has noted in *Entangled Edens*, many representations of the Amazon in Western media present the region as a virtually uninhabited space, populated by sparse numbers of indigenous peoples, and menaced by encroaching development (3–4). Such representations ignore the realities of large urban centers like Iquitos, Peru, and Manaus, Brazil, as well as the presence of people of many ethnic backgrounds who have called the Amazon home for generations.

In her book, Slater offers a detailed, thoughtful reading of how insiders and outsiders in the Amazon have understood their region.[5] Perhaps most importantly, she presents something lacking in many popular portrayals: a rich account of the human beings and communities who make the Amazon their home. Slater's study adds a voice to a growing chorus of writers and scholars who point out that environmental conservation, especially as it is often articulated in Europe and the United States, may be attempting to "protect nature from the very people who created it in its currently desired form" (Raffles and WinklerPrins 182). What these scholars point out is a history of human embeddedness in Amazonia—for better and for worse—that must be acknowledged in debates about the future of the region.

Like Slater, Souza, Sguiglia, and Sepúlveda also present a view of human communities in the Amazon, and they critique global and national plans for it. Their novels depict a host of different actors in two distinct regions (in Brazil and Ecuador), allowing readers a glimpse of competing worldviews, ideologies, and designs at work in the Amazon. Ultimately, their dark humor casts modernizing enterprises as exercises in absurdity with a high human and environmental toll.

Since the novels present contests for the Amazon, a general understanding of modern Amazonian history is essential in interpreting the texts. What has been the reality of the Amazon basin over generations? David Cleary broadly outlines the five phases of the environmental history of the Amazon: early human occupation, intensification of land management, depopulation following contact with Europeans, expansion of extractivism in the late nineteenth century, and rapid environmental transformation and destruction in the postwar period (68). Cleary points out that what nineteenth-century European naturalists actually saw and described as pristine forest was actually uncultivated, secondary growth of native vegetation (69). This was not virgin forest but rather the spread of vegetation into territories that had witnessed dramatic population decline. For generations after contact, indigenous populations succumbed to disease, relocation, and war. Early naturalists thus misinterpreted both history and landscape. Nonetheless, their writings, along with those of early Spanish and Portuguese conquerors, established enduring literary tropes in the West regarding the Amazonian world. Two of these are the Green Hell, a land of natural menace, and the lost Eden, a land of idyllic abundance.

How were Amazonian lands and waters really inhabited and experienced prior to European contact and shortly thereafter? To what extent did indigenous cultures shape Amazonian landscapes? The answer is elusive, especially if one reaches back into precontact times. Nevertheless, numerous studies suggest large human populations and well-organized societies inhabited various regions of the Amazon and not just the fertile floodplains.[6] Geographer William Denevan famously asserted in the "The Pristine Myth: The Landscape of the Americas in 1492" that the Americas were "well-populated rather than relatively empty lands in 1492" (370). Denevan gives his own estimate of the population of the New World at 53.9 million in 1492 (370).

As scholars like Denevan reconsidered population counts and habitation patterns, they also acknowledged a continuous human presence in areas long depicted as inhospitable to human settlement. Foremost among these areas is the Amazon. Aboriginal Amazonian peoples have a "deep history" of social organization and cultural development (Heckenberger xii).

Contact with Europeans brought about many changes in cultures and territories, many of which were mistaken for long-standing or permanent realities. In fact, anthropologist Michael J. Heckenberger remarks that until recently, scholars did not adequately appraise the extent of change wrought by colonialism: "The end of the fifteenth century marked a beginning in the Americas. . . . For many indigenous groups, it was the beginning of the end as European colonialism, notably the onslaught of imported diseases, resulted in rapid cultural disintegration and depopulation across broad regions" (1). By the time the earliest European and American naturalists arrived in the Amazon around 1750–1850, human populations had diminished significantly, and surviving ones had "become increasingly small and fugitive"

(Heckenberger 9). Moreover, remaining Amazonian populations changed their habitation and cultivation practices as they adapted to European, African, and mestizo presences. Now all these peoples, and more, form complex webs of human interactions within the landscapes of the region.

Contemporary media reports focus on many negative human impacts on nonhuman nature in the Amazon, but earlier populations of humans have sometimes had beneficial interactions with nonhuman nature. In fact, evidence indicates indigenous practices favored the maintenance and enhancement of biodiversity over time. Among the beneficial outcomes of particular cultural and agricultural practices, according to Denevan, was the creation of "humanized forest in which the kinds, numbers, and distributions of useful species are managed by human populations" (374).

Contemporary ethnographic research corroborates the historical possibility of an ecologically beneficial human presence (on biodiversity) in the Amazon. Most famous is the work of late ethno-ecologist Darrell Posey, who did field work with the Kayapo in Pará in the Brazilian Amazon. Posey "found that the Kayapo enrich rather than impoverish their rainforest habitat, by introducing food plants and fruit trees into their forest gardens. He and his co-researchers concluded that the great variety of the Amazon rainforests was due, in part, to the Amazonian forest tribes" (Girardet 57). Posey published these findings widely, and he championed the protection of indigenous intellectual property rights.

Why is work like that of Posey important? Popular depictions still suggest that regions like the Amazon were once uninhabited, pristine places where nature reigned supreme without human meddling. Often, misinformed environmentalists, particularly from developed nations, suggest that the best protection of these areas entails the removal and relocation of human inhabitants. The absence of humans might then allow flora and fauna to return to some "original" state. Such stances provoke ire and suspicion among many people living in the region, who suspect well-intentioned conservation efforts are just another attempt by outsiders to dispossess them of their lands.

Within nations that possess Amazonian territories, many people also wonder if there are ulterior motives at work behind conservation campaigns in the United States and Europe. J. Timmons Roberts and Nikki Demetria Thanos, for example, assert that "Brazilians are cynical after decades of 'gringo' interventions in the Amazon to extract rubber, minerals, lumber, and pharmaceuticals. Efforts to preserve the region are widely seen as attempts to prevent Brazil from gaining the superpower status it has long sought" (130). In light of the complicated history of imperialism, economic dependency, and intervention by Europe and the United States in Latin America, as well as exploitation by national elites of undeveloped territories, it is important that scholars, activists, and ecocritics from outside Amazonia work from well-founded scientific and cultural premises about the region. They also must take into consideration the long, complicated power struggles over the future of the peoples and places that constitute the Amazon.

The Amazon region has been at the center of competing claims at least since European contact. What is more, examples of dispossession and conquest riddle Amazonian history since then. During the colonial period, for example, the early mission system imposed by Portuguese and Spanish religious orders, like the Jesuits, offered native peoples protection from exploitation, but it also dislocated people from their places of origin and from one another. Cleary asserts that detribalized Indians from the missions "became the basis for the modern riverine mixed-race peasants" following the expulsion of the Jesuits in the 1750s and 1760s (88). Similarly, a century after the mission system ended, the extractivist economy, which reached its peak in the rubber trade in the late nineteenth century, produced another transformation in the orientation of the Amazonian communities. Trade during previous centuries had linked the lowlands of the Amazon to Andean highlands. As Latin America pursued export-oriented economies focused on trade with Europe and the United States, the South American regional economy turned inside out and became directed toward ports on the coast.

The South American rubber market collapsed in the early 1900s, but trade patterns were already established and new extractivist pressures came quickly. These included projects in mining, hydroelectric power, road construction, agrarian settlements, and large-scale, industrial agriculture. All transformed Amazonian reality as they opened up the region to new influxes of migrants. They also brought in their wake the rapid environmental destruction for which the Amazon is now famous and which Souza, Sguiglia, and Sepúlveda bring into focus in their works.

Mad Maria

Márcio Souza's *Mad Maria* takes as its subject the construction of a railroad deep in the Amazon in the early twentieth century. The Madeira-Mamoré railroad project had two purposes: to facilitate the export of rubber and to comply with the 1903 treaty obligations to Bolivia after the annexation of Acre.[7] Once completed, the railroad would allow rubber shipments to bypass the series of rapids on the Madeira River. Two hundred miles of track would be needed to circumvent the river falls, but a rail passage would expedite the safe transport of rubber to Atlantic ports, where it would be exported for booming automotive and electrical industries abroad.

In 1869 the Brazilian and Bolivian governments granted a concession to build the proposed railroad to George Church (Gauld 127). Faced with huge losses to human life and soaring costs, the company pulled out, and the endeavor collapsed in 1879. American entrepreneur Percival Farquhar bought a new concession for the railroad from a Brazilian speculator decades later, in 1906 (Gauld 126). Souza takes the intrigues surrounding Farquhar's involvement in Brazil, as well as the actual construction of the Madeira-Mamoré Railroad, as the subject of his novel.

Published in 1980, Souza's novel *Mad Maria* has been marginal for literary critics in Brazil and abroad. Brazil's TV Globo adapted the novel into a miniseries that aired in 2005, but to date, very little criticism has been published on *Mad Maria*. Most critical attention has instead gone to Souza's *Galvez, Imperador do Acre*, a historical novel set in nineteenth-century Acre state and published in 1983.

Mad Maria offers an intriguing glimpse into the heart of contested Amazonian territory along the Brazil-Bolivia border. Written and published during the dictatorship, it also gives a contemporary perspective on the collaborations of American investors and Brazilian elites. *Mad Maria* makes the argument that the complicity has a long history. Profit drives American investors, while ambition motivates Brazilian politicians. Neither shows concern for ordinary people or the extraordinary topography and environment of the Amazon.

Structurally, *Mad Maria* resembles a theatrical work or *telenovela*. The narrative skips from sexual romp to palace intrigue to labor camp brawl. There are numerous actors, including ambitious Brazilian political figures, cynical American company employees, a hapless and passionate Bolivian pianist, an armless Caripuna Indian, the conniving Percival Farquhar, and crowds of doomed, contracted laborers from all corners of the earth. The most important part of this novel, though, is the criticism of modernization born of collaborations between American investors and Latin American politicians.

The novel uses dark humor, parody, and burlesque to convey the message that the modernizing enterprise is an exercise in absurdity, fueled by hubris and greed. Critic Robert DiAntonio comments on *Galvez, Imperador do Acre* that "Márcio Souza's mordant humor underscores his penchant for exposing real socio-political problems" (269). The same could easily be said of *Mad Maria*. In *Mad Maria*, Souza draws attention to one of the most imposing business magnates in Brazilian history: Percival Farquhar, a man his biographer describes as having "the greatest land hunger of any man in the history of Latin America since the Incas" (Gauld 209).

Through a series of business enterprises in Brazil in the early twentieth century, Farquhar built railroads, developed ports, operated sawmills, encouraged U.S. colonization schemes, installed meatpacking plants, and reigned over a cattle kingdom of four million acres and 140,000 head of cattle in Mato Grosso state (Gauld 217).

Farquhar loomed large in Brazil for decades, and his attitudes, as well as those of his biographer, are indicative of the political and economic machinations of U.S. investors in Latin America. Neocolonialist projects like Farquhar's paired foreign capital with the ambitions of Brazilian politicians and speculators (largely located in coastal cities like São Paulo and Rio) to transform what were for them remote territories inhabited by Others, in this case, indigenous and mixed-raced riverine people in the Amazon region.

Later, the language of business was wedded to the rhetoric of the Cold War. For example, in the foreword to Farquhar's biography in 1964, Ronald Hilton wrote that Brazil "is potentially a second Cuba. . . . At present the trend in Brazil is to reject Farquhar's thesis of development under U.S. guidance and with the United States holding key levers in the national economy" (Hilton xi). When the financial elite of the United States held "key levers" in national economies other than their own, it meant they held the power to impose upon the human and nonhuman worlds of Latin America the same models of "progress" that radically transformed the environment of the

American West. Superimposed on Latin American landscapes and peoples, such activity marginalized local knowledge of ecosystems and altered a long history of human interaction with the natural world in the Amazon.

In *Mad Maria*, Souza paints infrastructure development in the Amazon as a profit-driven scheme forced on territories for which such development was imminently unsuitable. In 1980, as dictatorships in the Americas embraced infrastructure investments in the service of export-led growth, Souza lifts a dissenting voice from within the Amazon region. Souza is very cognizant of his identity as a critic, and the political overtones of his art forced him to leave Manaus during the dictatorship. More recently, he has remarked of the comparative neglect of the literary world toward Amazonian writers that "maybe we need more deforestation here to get some attention" (Rohter).

The social, political, and environmental backdrop of *Mad Maria* at the moment of its publication in 1980 has interesting parallels with Farquhar's era in Brazil. By 1980, the period of military dictatorship that began in 1964 was well into its second decade and reaping a debt crisis brought on by development projects financed by international lending institutions. From 1968 to 1974, Brazil had experienced a period of rapid economic growth ("the Brazilian miracle"), fueled by transnational investments that drove the production of durable consumer goods. During that time, Brazil instituted modernization projects in agriculture that were intended to increase agricultural exports and offset the costs of importing petroleum (De Souza, et al. 15–16). After 1974, rising oil prices hit the country hard, growth diminished, and large state enterprises looked to international banks to finance their projects. By the 1980s, the consequence was a massive debt to international banks (De Souza, et al. 19).

In the midst of the boom, the military dictatorship announced plans to construct the Transamazon Highway, one of the grandest infrastructure projects in the Americas. Begun in 1970, the 3,300-mile Transamazon Highway was opened by 1975 as a "two-lane dirt road" (Smith 755) at a cost given at $500 million in a 1981 article (Smith 760). But what was the source of the capital? According to Oxfam's Patricia Feeney, "much of the funding for the rapid development and colonisation [sic] of the Amazon was provided by the World Bank, which during the 1970s and 1980s implemented a number of large projects, ranging from power generation schemes to integrated rural development projects" (3). Luiz C. Barbosa points out that "the World Bank and the Inter-American Development Bank contributed to the [Transamazon] project by lending US$400 million to the Brazilian National Highway Department" (321).

The environmental impact of the Transamazon Highway and other infrastructure projects has been immense. According to Emilio Moran, in 1975 a total of 3 million hectares of Brazilian Amazonia had been cleared; by 1980, 12.5 million hectares had been cleared and the rate of deforestation was accelerating rapidly (1). Nigel J. H. Smith describes the Transamazon Highway as a monumental failure, another government-led development scheme drawn up "with little or no understanding of the ecological and cultural conditions of settlement area" (760). The Transamazon did not accomplish colonization, development, or extractive objectives, and its cost contributed to a massive external debt (Smith 760). However, it did push the capitalist frontier deep into Amazonia, creating a clash between what Barbosa calls "people of the forest" (indigenous people, rubber tappers) and agents of international capital (318).

So what is Souza's view of the vast Amazonian natural world? It is not an ecocentric depiction, and contemporary environmental rhetoric does not color his portrayal of the Amazon. Indeed, there was little international media attention to Amazonian issues in the 1970s; most concern came from scientists who published in specialized journals like *Science* (Moran 9). In *Mad Maria*, Souza levies a discourse of nature to condemn a corrupt government in collusion with foreign investors. Publishing in the midst of dictatorship in the Southern Cone, with neoliberal models taking root all over Latin America, Souza fictionalizes an earlier period, from 1907–1913, with parallels to the Brazil in the 1970s and 1980s.

Souza does not romanticize the area along the Madeira River in the upper Amazon. Indeed, he appears to trope the region as a Green Hell, offering a vision of mud, torrential rain, stinging insects, and rushing rivers. Similarly, gender and racial stereotypes abound: men are consumed with a desire for power, women are sexualized, indigenous people are childlike or impenetrably other. However, the exaggerated, parodying tone of the novel means these depictions cannot be taken as straightforward representations.

The opening passage of Portuguese-language original offers a guide to interpretation:

Quase todo neste livro bem podia ter acontecido como vai descrito. No que se refere à construção da ferrovia há muito de verdadeiro. Quanto à política das altas esferas, também. E aquilo que o leitor julgar familiar, nã estará enganado, o capitalismo não tem vergonha de se repetir (11).
[Almost everything in this book could have happened the way it's told. Insofar as it details the building of the railroad, I have tried to be meticulous—likewise with the politics of the powers that be. And wherever the reader judges something to be familiar, he is probably not mistaken. Capitalism seldom has been ashamed to repeat itself (unpaginated foreword).]

Interestingly, the Portuguese original explains that "this book is nothing more than a novel" and issues a command to the reader to "pay attention" (11). The English version follows the opening paragraph immediately with the famous quote attributed to American President Franklin D. Roosevelt about Nicaraguan dictator Anastasio Somoza, that "he may be a son of a bitch, but he's our son of a bitch." Only then does the English version reproduce the same admonition and command, all of which appears as a foreword instead of as an introductory paragraph in the first chapter.

Narratological tools are essential in decoding representations and deciphering meanings in the text. For example, Souza makes abundant use of focalization, so many passages about the Brazilian Amazon are filtered through the eyes of less-than-reliable characters. The characters themselves are exaggerated into parodies: the ambitious investor, the gruff engineer, the swooning damsel in distress.

The first glimpse *Mad Maria* offers of the Amazon region comes on the opening page of the novel. It is filtered through the eyes of the young, privileged, and naïve American medical doctor working as a company physician at the Madeira-Mamoré railroad construction camp:

Finnegan não sabia que os escorpiões começavam a aparecer no começo do verão.
E o que era o verão naquela terra, afinal?
. . .
Era o primeiro verão que Finnegan estava passando ali e começava a aprender sozinho a lidar com os escorpiões. Ninguém tinha lhe falado de escorpiões (11).

[Finnegan had no idea that scorpions would appear with the onset of summer. As a matter of fact, what in blazes did summer mean in a place like this anyway? . . . This would be the first summer Finnegan was to spend here and he quickly taught himself to make war on scorpions. Why had no one even bothered to warn him? (3)]

The opening parodies the plot line of "man versus nature," as it relates a futile, risible battle the diligent newcomer wages against the small, stinging invaders of the ordered company infirmary. It also foreshadows his later violence against the workers. The focalization conveys a subtler message, though, since Finnegan's attitude toward the scorpions also illuminates the perspective of others. The fact that other people did not "bother" to warn him belies the fact that for those who know the region, the presence of scorpions during rainy season is routine.

At the heart of Souza's novel is the juxtaposition of two visions of development in the Amazon region during the early twentieth-century period of national expansion and infrastructure development. The first is the vision proffered by men like Farquhar to the Brazilian government officials from whom he courted concessions, friendly investment terms, and legal favors. In this view, development of the Amazon region would tame the forces of nature for civilization, bringing export commodities to market and prosperity to investors. The vision was persuasive; according to William R. Summerhill, "from only a few hundred kilometers of track in the 1860s the system expanded to more than 24,000 kilometers in 1913" (74).

The other vision of development in the Amazon is more critical: the installation of track cost countless human lives, disrupted indigenous cultures, and enriched a few powerful individuals. Souza structures the novel so as to present the world of mud, toil, and death of company laborers first. He contrasts these images with chapters set in interior spaces in Rio, in luxurious dwellings and the national palace where Farquhar's political dealings take place.

Ultimately, the two visions converge in crucial chapters near the conclusion of the novel. In Chapter 16, Souza paints a picture of the town of Santo Antônio, a noncompany town formed by displaced indigenous people and migrants to the Amazon region. In Chapter 19, Souza exaggerates the stereotypical, gala visit by politicians to tour a new business venture.

Souza is at his sardonic best in these two chapters. A chapter set in Rio reveals that a Rio journalist who cannot be bought has made allegations about Farquhar's Amazonian projects. Next, the reader sees Farquhar's plans for a publicity stunt, in which a delegation of Brazilian politicians will travel to the company town of Porto Velho to see progress with their own eyes.

The parody of the gala visit includes attractive nurses who attend high-ranking politicians who fall ill, speakers who ply attendees with lofty rhetoric, and housing in luxurious accommodations filled with red carpets and period pieces shipped in from Manaus. The pièce de résistance of the Porto Velho visit is a piano performance by the armless indigenous man dubbed Joe Caripuna by company employees. Joe Caripuna is the figure on which Souza plots the trajectory of the life of Brazil's indigenous peoples.

Early in the novel, a passage focalized through Joe presents readers with the history of the man that haunts the railroad construction site in search of food:

Os brancos civilizados nào gostavam de acordos e preferiam roubar as mulheres e atirar nos homens. Um dia tentaram roubar a sua pequena tacuatepes mas ela nào queria ir e se debateu e gritou com tanta fúria que um civilizado abriu ela com um golpe que saía do pescoço e acabava entre as pernas dela. Ele a encontrou morta dentro de um tacho de fazer beiju, boiando no sangue já escuro e as pernas escancaradas onde as moscas voavam. Naquela época a maloca quase nào tinha mais nenhuma família, muitos tinham se mudado para além da serra dos pacaás-novos, ou estavam mortos, ou viviam junto dos civilizados trabalhando como seringueiros ou bebendo cachaça em Santo Antônio (68).

[The civilized, though, rejected understanding and preferred stealing women and shooting men. On one occasion they had even tried to steal his young Tacuatepe, but she had refused to go and had struggled and screamed with such fury that a civilized opened her up with a blow that began at her chest bone and finished between her legs. He found her dead inside an earthenware vessel for making tapioca, floating in the already darkened blood and her legs wide open where flies were circling and landing. At that time, their maloca had had hardly any families left; most of them had fled to beyond the mountain range of the Pacáa-Novo, or were already dead, or living among the civilized, working as seringueiros and getting drunk on cachaça in Santo Antônio (71, italics in English translation).]

Joe has gone from a member of an integrated social unit to a refugee. Shortly after, he will become disabled.

The gruesome depiction of violence against his wife stands as a metaphor for the rape of the Amazon region and her human inhabitants; she floats in her own sustenance, and flies feed on her genitalia, symbols of fertility and pleasure. Like his companion, Joe, too, is violated when company employees administer their own justice after he steals from their tents. He loses not his life but his arms, limbs crucial for both work and self-defense.

Unfortunately, there is no real humanization of the indigenous Other in *Mad Maria*, as there was in the Argentine novels by Iparraguirre, Demitrópulos, and Belgrano Rawson. Joe is simply a vehicle by which Souza criticizes Farquhar's enterprise because an armless Joe Caripuna has no place in a productive Amazon. As a consequence, Farquhar ships Joe out to Rio so he can pay his debt for medical care. When Rio society shuns his piano performances by foot, Farquhar contracts Joe to American curiosity show founder P. T. Barnum, for whom Joe works until he dies of syphilis. The trajectory of Joe's life becomes a metaphor for the fate of the indigenous population of the Brazilian Amazon. Their lives and cultures do not fit into development projects that see them as obstacles and curiosities. In the world of *Mad Maria*, they either die or become parodies of themselves as a consequence.

The depiction of the Porto Velho state visit, and the epigraph on the life of Joe Caripuna, are stinging indictments of Amazonian schemes. They are made all the worse because they come on the heels of Souza's darkest chapter. The town of Santo Antônio carries the name of the patron saint of lost things, and it is a place of the lost and the damned. Souza delivers a view of the aftermath of development by means of the escape of Collier and Finnegan from the Porto Velho hospital, where they were recovering from earlier injury and illness. Bored, they steal a canoe and emerge from a dark, moonlit night into a world that is an Amazonian hell. Streets are riddled with potholes, plazas are bathed in cattle blood, and indigenous women are sex workers. Seringueiros (rubber tappers), flush with cash on payday, carouse there, and rubber barons buy expensive liquor with new wealth. There are no children anywhere because they all die from the miserable circumstances of their birth.

The Santo Antônio chapter delivers the ideological message of the novel. Souza pronounces it in the voice of railroad engineer Stephan Collier, an itinerant railroad man brought out of retirement by Farquhar. Collier is the most appealing figure in the novel, and like many other protagonists in works of "ecological imagination," he is a hybrid figure homeless in the modern world. An Englishman by heritage, Collier is a Civil War veteran from the Confederate South. After losing his family in the war, he is educated in England and afterwards, works in the booming international railway enterprise. Displaced by war and also by choice, Collier is a keen observer and critic, as well as participant, in the modernizing enterprise. This dialogue between Finnegan and Collier delivers Collier's critique:

—Você é um engenheiro, Collier, não um policial.

—É a mesma coisa!

—Não concordo. Você e eu trabalhamos pelo progresso.

—Um caralho! Quer saber o que significa para mim o progresso? Uma política de ladròes enganando países inteiros. Birmânia, Índia, África, Austrália, os nossos alvos.

—Mais estamos deixando a nossa marca.

—É claro que estamos deixando a nossa contribuiçaò. Ao lado da cadeia de tijolos, está a escola para formar funcionários nativos subalternos. Nós nào esquecemos nem de ensinar os jovens nativos o futebol. E aprendem a beber uísque, principalmente a beber uísque. . . . E enchemos a cara enquanto enriquecemos, enquanto destruímos tudo, enquanto espalhamos os nossos própios vícios (258).

["You're supposed to be an engineer, Collier, not a police officer." "Down here, it all amounts to the same thing." "I disagree. You and I are here to work for progress." "Sheer rubbish! You want to know what progress is? A pack of thieves in politician's clothes humbugging an entire nation: Burma, India, Africa, Australia—those are the targets." "So, we're leaving our mark on the world!" "Yes, we make our little contributions. Next to the brick jail is the wooden schoolhouse to train native bureaucrats in subservience. And don't forget to teach the young babus a little soccer. And when they get older, to drink whiskey—especially to drink whiskey. . . . All this, while we fill our faces and grow rich! While we infect the world with our own vices, destroying everything!" (290)]

This denunciation of neocolonialist collaborations between national elites ("a pack of thieves in politician's clothes") and foreign capitalists is the heart of the novel. Education, sport, and leisure, under their auspices, serve an unholy alliance that "destroys everything."

The remainder of the novel reinforces this message. The natural world of the Brazilian Amazon and its human communities fall at the hands of ambitious individuals with no knowledge of local terrain or traditions. The story is an environmental and cultural tragedy, levied by Souza to critique the project of modernization.

So where is the absurdity in this tale? Souza delivers the news flatly in the last chapter when the burlesque narrative ends abruptly. Ordered like a philosophical proof, short sentences deliver the final blow:

> No dia 7 de setembro de 1912, à revelia do governo brasileiro, foi inaugurada a estrada de ferro Madeira-Mamoré.
> Em 1912, a borracha de Amazônia tinha perdido o monopólio internacional para as plantaçóes inglesas na Asia.
> Em 1912, a estrada de ferro Madeira-Mamoré, aparentemente, començava a deixar de ter sentido (339–40).
> [On the seventh of September 1912, without the knowledge or approval of the Brazilian government, the Madeira-Mamoré Railroad was finally inaugurated. By 1912, Amazonian rubber had already lost its international monopoly to the English plantations in Asia. By 1912, the Madeira-Mamoré Railroad had, in effect, ceased to have any purpose for its existence (385).]

The mocking tone of the novel dissolves, and the language becomes understated and direct. History, in the form of global geopolitics, renders the final blow for the Madeira-Mamoré Railroad. And for readers who would look at the Amazon region in 1980 and wonder about parallels, Souza might point them again to the foreword of *Mad Maria*: "E aquilo que o leitor julgar familiar, nã estará enganado, o capitalismo não tem vergonha de se repetir" (11) [And wherever the reader judges something to be familiar, he is probably not mistaken. Capitalism has seldom been ashamed to repeat itself (unpaginated foreword).]

Fordlandia

Published nearly two decades after *Mad Maria*, Eduardo Sguiglia's *Fordlandia* levies the mythos of the Amazon to critique yet another neocolonial adventure in Brazil. Environment, race, and politics again figure prominently in Sguiglia's depiction of the Amazon. His ecological imagination is also very anthropocentric, serving to critique foreign investment schemes prompted by global competitition for rubber.

Sguiglia recuperates a sense of place, but he does so following the conventions of a travel narrative. In this case, the travel narrative reaffirms the indomitability of the Amazon. In the journey, the traveler-narrator gains a sense of place, but he uses it to condemn neocolonialist powers instead of to celebrate the nonhuman natural world.

The resilience of the Amazon, as metonymy for Latin America, was an attractive concept when Sguiglia published his novel. By 1997, Latin American economies had been deeply structured by neoliberal agendas pushed through since the 1980s. With structural adjustments in place as preconditions for lending by international monetary institutions, the middle and lower classes of most Latin American countries paid a heavy toll. Cuts in state spending eliminated middle-class jobs and important social services. The privatization of industries sparked accusations of plunder by politicians and fueled nationalist critiques of globalization. Throughout the 1990s, the perception grew among many Latin Americans that foreign companies wanted access not just to markets but also to society and its structures. Zapatista leader Subcomandante Marcos became the most famous voice of dissent and levied strong-worded denunciations of globalization in essays like "The Fourth World War Has Begun."[8]

Written in the midst of such reality, *Fordlandia* offers a depiction of environmental change brought about by the failed Amazonian experiment, decades earlier, by capitalist par excellence Henry Ford. Like Souza's novel, *Fordlandia* takes place within the Brazilian Amazon. Sguiglia, like Souza, splits the narrative between Amazonian territories and interior spaces elsewhere, in this case, the Ford mansion and headquarters in Michigan.

The Amazon region in question is the area along the banks of the Tapajós River near Santarem, where the northward-flowing Tapajós joins the Amazon. Historically, the Munduruku native peoples living near the Tapajós and Cururu rivers dominated other indigenous groups. This dynamic began to change when the rubber boom of the nineteenth century brought increasing waves of outsiders to Munduruku territories there. More recently, the area around Santarem has seen a boom in soy production, stimulated in part by the construction of a processing and export facility by U.S.-based firm Cargill.

Sguiglia's focus is the rubber plantation and company town known as Fordlandia, founded on a vast tract of land acquired from the state of Pará (Nevins and Hill 233–34). Fordlandia was one of two communities Henry Ford planned in the Brazilian Amazon, the other, established after Fordlandia failed, was called Belterra. The projects grew out of a campaign spearheaded by Harvey Firestone to circumvent British control of the rubber market. The result of extensive study for site selection culminated in 1927, when Detroit attorneys O. Z. Ida and W. L. Reeves Blakely "negotiated an agreement granting the automaker 2.5 million acres deep in the Brazilian Amazon, police protection and duty-free entry of all Ford equipment and supplies" (Dempsey 2). The site had been chosen on the basis of a U.S. government report made by University of Michigan botanist Carl LaRue. It was soon planted with 1.4 million rubber trees (*Hevea brasiliensis*) (Dempsey 2).

Sguiglia's fictionalized story of Fordlandia takes historical facts relating to Fordlandia and Belterra and conflates them into one narrative. Ford's project failed miserably, and the story was largely ignored in Brazil and the rest of Latin America until Sguiglia appropriated it for his own ends.[9] Sguiglia accurately represents the motives for the settlement and the spirit of the enterprise. Driven by competition and profit, Ford sought to harness the productive capacity of the tropics to enrich an automotive empire. In this mentality, "useless" parts of the forest would fall to make way for an organized, monoculture plantation of evenly spaced rubber trees. Efficiency would triumph over disorder, Ford would get richer, and the Amazon would become a productive landscape. Unfortunately for Ford, results did not follow as expected. Monoculture cultivation of rubber trees provided the perfect environment for blight, mites, and disease that are controlled in mixed-species forest. The rules and order of Ford in Michigan created catastrophe when they were applied with little knowledge of local ecology or local people.

Sguiglia's novel recuperates the memory of Fordlandia to depict it as a colossal failure. Enriched by a fictional cast of characters, the novel deviates from historical records in significant ways. One such episode is the completely fictional visit by Henry Ford that occurs at the end of the narrative.

In narratological terms, Sguiglia's novel is complicated. Most of what readers learn comes through the filter of the homodiegetic narrator, Horacio. Horacio is a young Buenos Aires resident hired by Henry Ford's representatives to oversee Fordlandia operations, particularly labor recruitment and retention. He goes unnamed until the last pages of the novel when Ford himself utters his name in a business meeting.

Other significant characters come from outside the Amazon, too, and Sguiglia definitely privileges their lives and stories over those of Amazonian residents. Caroline, for example, is a Canadian anthropologist employed by Ford. With a genuine interest in residents of the Amazon, she compiles careful ethnographic studies that "dejaban traslucir la probabilidad de que las ideas de Henry Ford fracasaran allí" (67) [hinted at the possibility that Henry Ford's ideas would fail there (53)]. In Caroline, Sguiglia creates a character motivated by intellectual curiosity who enjoys the respect of the people she studies. Nevertheless, her work fits into an uncomfortable economic reality. She only receives funding to carry out her ethnographic research because her employer wants to use it to procure laborers for his plantation. Caroline is the figure Sguiglia uses to bring into the foreground the complicity of social science research in human and environmental exploitation.

Along with Caroline and the narrator are characters like Jack, a troubled ex-convict Ford employee who befriends the narrator. Rowwe, Ford's representative in Fordlandia, is a petty and scheming lackey interested principally in ascending the ranks of the company. Theo is a Catholic priest at the peripheries of the Ford project, one who colludes and collaborates when it is in his own interest to do so. Through Theo, quick to ally himself with the powerful, the Catholic Church appears as an interested party in the drama of the Amazon region.

The only two important characters drawn from the local population are Roque and Enéas, assistants to Horacio. Sguiglia uses them to register the history of the Amazon in their ethnic make-up and their attitudes toward one another. Enéas is the narrator's assistant, and he is "en la opinión de la gente, . . . un caboclo con antepasados indígenas y africanos. Para él, sin embargo, los indígenas era los verdaderos caboclos. Que no lo llamara caboclo fue lo único que me pidió a poco de conocernos" (51–52) [People thought that Enéas was a caboclo with indigenous and African ancestors. For him, however, the natives were the real caboclos. The only thing that he asked me, shortly after we met, was that I not call him caboclo (38)].[10] When Horacio begins a river journey to recruit laborers, Enéas, who loathes manual labor, offers to find another man to accompany them. Roque is that man.

> Era mulato y más bajo que Enéas. Parecía joven—casi un muchacho—y había nacido en una antigua villa al lado del Tapajós. . . . Roque, a diferencia de Enéas, caminaba rápido y pisaba la tierra con orgullo. Enéas poseía la facultad de orientarse en los ríos y en la selva, aun bajo circunstancias difíciles. Roque, en cambio, demostró, a lo largo de todas las incursiones, ser buen cocinero y un conversador insuperable. Sabía todas las leyendas y mitos del Amazonas (60).
> [Roque was a mulatto and shorter than Enéas. He looked young—almost like a boy—and was born in an old village on the banks of the Tapajós. . . . Unlike Enéas, Roque walked quickly and proudly. Enéas had the ability to find his way around the rivers and jungle, even under difficult circumstances. Roque, on the other hand, proved throughout the trip to be a good cook and an unbeatable conversationalist. He knew all the myths and legends of the Amazon (47).]

These characters offer commentary on local life for the narrator, and in this way, Sguiglia imparts information about the social, cultural, and ecological worlds of the Amazon.

Sguiglia draws Ford's ideologies into the narrative by interspersing brief chapters focused on Michigan with the main narrative about the Amazonian drama. The juxtaposition of chapters plays up the philosophical, cultural, and environmental differences between the two spaces. Ford applies the same reasoning about productivity, organization, and discipline that worked in Michigan to the territories and people he employs in the Amazon. The Michigan chapters highlight Ford's racist and moralist ideologies behind the work ethic. Combined with Horacio's sharp commentary, the juxtaposition of chapters reveals the incongruousness of Ford's ideologies in the Amazon, at the same time that it attacks them at their racist roots.

Horacio is the hybrid figure who navigates between company rhetoric and Amazonian realities. Like Iparraguirre's Guevara, he is key to delivering the ideological message of the text. Horacio is a Latin American insider but Amazonian outsider, professional but subaltern, critic but participant in a modernizing enterprise he disdains. Horacio uses this position of being in between cultures to wield authority. Similarly, Sguiglia privileges Horacio's voice, using it to control reader's perceptions of Fordlandia. For example, when Ford's Sociology Department produces guides to healthy, moral living, Rowwe has these translated into Portuguese and Tupí. The purpose of the guides is "mejorar a los hombres del mismo modo que la cadena de montaje instalada en sus plantas le permitía mejorar los autos" (66) [to improve men in the same way that the assembly line installed in his plants had allowed him to improve cars (52)]. After a long, descriptive paragraph detailing the contents of the pamphlets, the narrator states his flat refusal to distribute them. Passages like this highlight the insider knowledge of the narrator, and they subvert the confident, capitalist rhetoric of Henry Ford and his representatives.

Sguiglia constantly portrays Horacio as a more astute reader of human nature than Ford or his representatives. However, he also crafts a narrative in which Horacio himself is blinded to certain realities, particularly those regarding the Amazonian landscape. It is essential for an ecocritical reading to navigate this narratological maneuvering on Sguiglia's part.

Horacio at first only perceives the region through the tropes of literature, and Sguiglia makes this filter transparent on the first page of the novel: "Recordé que durante años, prácticamente desde mi niñez hasta la juventud, tuve pasión por los libros de viajes. Y también que, al cabo del último renglón de la última página, me quedaba pensando (en vano) sobre cuánto de verdad y cuánto de ficción encerraban los relatos" (7–8) [I recalled how for years, practically from the time I was a child until I was a young man, I was passionate about travel books and how after the last line of the last page I would wonder (in vain) how much truth and how much fiction those tales contained (2)]. Sguiglia deftly uses dialogue to reveal the limitations of the narrator and provide readers with perspectives of other characters who read the landscape differently. Through their filters, the Amazonian world of non-human nature and human communities becomes visible, and it is never fixed or stable.

A self-confident urbanite, Horacio has little appreciation for the Amazon at the outset of his adventure. The region represents only an escape from a troubled past of debts and discontentment. In fact, he is initially hostile and defiant in the face of the green monotony of the Amazon.

As he experiences Amazonian realities, Horacio's annoyance gradually evolves into a respect for place. He learns the region does not bend to the will of powerful men: "También advertiría, como Plinio veinte siglos atrás, que un suelo adornado por miles de árboles altos y hermosos no siempre es favorable, excepto, claro está, para esos árboles" (8) [I would also realize, as did Pliny twenty centuries before, that a land covered with thousands of tall, beautiful trees is not always a good thing, except, of course, for the trees (2)]. This quote in the opening chapter references an outsider's filter, in this case the work of Roman naturalist and historian Pliny the Elder. Like other travelers before him, classical literary works provide the lens by which the narrator processes the lessons of the Amazon. Eventually, Horacio expresses not an affinity, but an allegiance, with the Amazon, simply because it defeats the arrogant enterprise by Henry Ford.

As the novel opens, Horacio initially likes the looks of Fordlandia as a break in the unbroken landscape of trees. Despite this, he sees in Fordlandia no indication of the imposing reputation of Ford and his empire:

> Un rato más tarde divisé Fordlandia. Nada de lo que se ofrecía a la mirada era monumental ni insigne; sin embargo, vista desde el río la villa era una pausa agradable a la inacabable sucesión de lo verde cerrado. Recostada sobre la margen izquierda aparecía como un refugio amplio y seductor (31).
> [A little while later I spotted Fordlandia. Nothing visible to the eye was either monumental or noteworthy; however, seen from the river the village was a pleasant interruption in the unending succession of dense greenery. Set on the left bank, it looked like a vast and seductive refuge (21).]

The initial sketches of nonhuman, Amazonian nature replicate Western narrative tropes regarding the Amazon. However, they come in the voice of only a somewhat reliable narrator, and they must be interpreted in light of later episodes in the novel.

The most important episode of the novel is Horacio's journey upriver. Sguiglia makes this journey the centerpiece of the plot; it is the means by which Horacio comes to consciousness about social and environmental realities. The river journey signals the importance of waterways, and it also aligns the text with travel narratives and novels like Joseph Conrad's *Heart of Darkness*. The purpose of Horacio's journey upriver is to recruit laborers for the rubber plantation from among indigenous groups in the region. For readers, it is also the means by which Sguiglia presents the socioecological realities of the Amazon that ultimately bring Ford's neocolonialist plans to a halt.

Early in the journey, Horacio identifies with the civilizing rhetoric of outsiders, and he repeats tropes of the Amazon as primeval landscape:

Al cabo de un día y medio de navegación, nos acercamos a una aldea de la tribu munduruku. Los ríos y los arroyos fluían desiertos y tuve la impresión de que al remontarlos regresábamos a los orígenes del mundo. Pero regresábamos como precursores del cambio, del progreso. Eramos el fragmento de una epopeya, la epopeya de la conquista del reino de los árboles. La idea me fascinó y deambuló por mi cabeza algún tiempo (127).

[After traveling for a day and a half we reached one of the villages of the Munduruku tribe. The rivers and streams were deserted, and as we moved along I felt as if we were returning to the origins of the world. But we were returning as precursors of change, of progress. We were an excerpt in an epic poem, the epic of the conquest of the kingdom of the trees. The idea fascinated me, and it lingered in my mind for a while (109).]

Here the narrator repeats a common misconception about the Amazon, that is, that "its people are archaic" (Heckenberger 7). He sees himself as part of a grand enterprise that will initiate progress. Sguiglia here aligns Horacio's hubris with that of Henry Ford, and he records the appeal of civilizing rhetoric for outsiders eager to see themselves in a messianic role in the Amazon.

Sguiglia proceeds to deconstruct this rhetoric piece by piece during passages about the journey and Horacio's time in Fordlandia immediately after it. Ford embodies the narrative of progress, of machine over nature, and Sguiglia paints him as both a racist and, at least when it comes to the Amazon, a fool. As for Horacio, the inclusion of dialogue with Amazonian insiders reveals that his own assertions about both the natural world and human societies in the Amazon are simplistic and incorrect.

The turning point for Horacio, in terms of his own coming to consciousness, is his disastrous foray into the forest. On the long return to Fordlandia, the narrator says he is "harto de las supersticiones, de los hechiceros, de las historias de Roque, de los mitos, de que cualquier impulso hacia la selva estuviera atajado por el miedo" (146) [fed up with superstitions, witch doctors, Roque's stories, the myths, and with the fact that every impulse toward that jungle was thwarted by fear (125)]. Horacio rebels against a landscape interpreted, reinterpreted, personified, and protected by oral traditions related by his assistants. With that, he takes up arms against the rumors and enters the forest, ostentibly to hunt. The vegetation rises up against him, or so he thinks; he makes it less than four hundred meters from the riverbank camp before succumbing to insects and his own fears.

The passage is crucial for an ecocritical interpretation of *Fordlandia*. First, it reveals the extent of human fears about the more-than-human world with which they are not familiar. Second, because it depicts the protagonist in an extremely unflattering light, it is key to interpreting the narrator's voice and hence the entire novel. Upon setting out, the narrator asks "Si no podíamos dominar a esas especies inferiores etaba seguro que ellas nos dominarían a nosotros. ¿Y qué había allí adentro sino especies inferiores y miles de árboles paralizados, inmóviles, condenados a esperar a que la mano del hombre, más tarde o más temprano, los hiriera o los derribara?" (146) [If we could not overpower those inferior species, I was certain that they would overpower us. And what was there, inside the jungle, but inferior species and thousands of motionless trees, condemned to wait for the hand of man, sooner or later, to hurt them or knock them down? (126)] Disdainful and ignorant of Amazonian realities, and decidedly anthropocentric in his thinking, his foray becomes a self-fulfilling prophecy of a Green Hell rising up against him. In a matter of minutes, the narrator finds himself in the forest, surrounded by insects, sweating, afraid, and shooting at spiderwebs. He loses consciouness just as he sees a snake approaching the place where he has fallen.

Finally, Sguiglia makes the importance of the passage clear, after the return of the expedition to Fordlandia. Recovering in the company hospital, the protagonist learns from Roque about the admiration expressed by some of the Amazonian people he encountered on the journey. He attributes their respect to his defiant entrance into the "jungle." Instead, he learns an im-

portant and decisive lesson. The local people do not admire him because
he entered the jungle but because he confronted the powerful landholders:

> —Comprendo. ¿Y por qué crees que aquellos me miraron así?
>
> . . .
>
> —Su valor es considerado, mi blanco.
>
> —¿Por entrar en la selva?
>
> —No, ellos conocen bien la selva. Usted está considerado por su valor
> con los puesteros de la zona, eso sí (167).
>
> [—I see. And why do you think that they look at me that way?
>
> —Your courage is respected, mi blanco.
>
> —For going into the jungle?
>
> —No, they know the jungle well. You are respected for your courage
> with the post owners in the area. (145–46)]

One of the first episodes in the river journey was Horacio's hotheaded
confrontation of a rubber baron loathe to let his laborers go to Fordlandia
(or even hear of it). The hospital conversation, coming immediately after
Horacio's hunting fiasco, forces Horacio (and readers) to revise their view of
the landscape. As a city dweller, the narrator thinks the forest is dangerous,
fearsome, and in need of domination. Roque points out matter-of-factly that
local residents know the forest well. They do not fear the Amazon but rather
local strongmen who use violence and money to control their lives.

The narrator articulates no epiphany in the text, but by the time he leaves
Fordlandia, his actions and language reveal a new understanding of the
rhythms of the Tapajós landscape. The chapter that catalogues the unravel-
ing of the project in Fordlandia begins with passages, narrated by the protagonist, that detail seasonal changes that begin with the arrival of summer
(dry season). The opening paragraphs convey a change in perspective on the
part of the narrator. After a year and a half in the Amazon, Horacio literally
begins to draw closer to the natural world ("long walks" along the banks) and
abandons the obviously futile, disciplined work routine imposed by Ford:

> Aproveché los primeros días de verano para hacer largas caminatas por
> la villa y por las orillas del Tapajós. . . . En breve tiempo el sol se encargó
> de secar los lodazales; brotaron, aquí y allá, miríadas de mariposas y,

al atardecer, volvieron a escucharse las estridencias de las cigarras. En el aire y en el paisaje se percibía que la vida se había renovado. En la plantación, sin embargo, las cosas iban muy mal (227).
[I took advantage of those first days of summer to take long walks through the village and along the banks of the Tapajós. . . . In just a short time the sun dried up the muddy spots; myriads of butterflies appeared, and at dusk the shrill song of cicadas was heard once again. In the air and in the countryside there was a sense of renewed life. On the plantation, however, things were not going well (199–200).]

The passage records Horacio's growing body of local knowledge about weather and fauna, particularly his observations of the migratory patterns of the birds.

The passage also underlines the most salient fact about Amazonian reality: that Fordlandia is out of step with the local landscape. The rest of the chapter details the blight on the rubber trees that ultimately leads to the abandonment of Fordlandia. The discourse of nature serves to underline the absurdity of modernizing plans.

Sguiglia structures the novel so that the last vision of Fordlandia comes after a visit by Ford himself to the settlement. Ford's assertions about man's triumphs over all the jungles of the world ring hollow in the face of the reality around him. The narrator's last vision of Fordlandia reinforces the futility of the project:

La plantación nunca más se restableció. Si los hongos la mantenían en trance, las langostas la terminaron de rematar en un par de jornadas. La naturaleza de la villa se marchitó rápido y los edificios, cubiertos sin demora por la selva, sólo mostraron, con el tiempo, sombras de viejas amarguras (274).
[The Plantation never recovered. The locusts finished off in a couple of days what the fungus had started. The natural beauty in the village quickly withered, and the buildings, which were soon swallowed up by the jungle, with time looked like mere shadows of old failures (243–44).]

As the novel approaches its conclusion, it is clear that Amazonian nature is a metaphor for Latin America, which reclaims its own territory.

Sguiglia's ecological imagination in *Fordlandia* is problematic. The nonhuman world as a real entity is reduced to a metaphor about human dramas. Furthermore, the metaphor emphasizes the resilience of the flora and fauna of Amazonia, one in which readers might become too confident given the natural limits of ecosystems to restore and regenerate themselves.

Nevertheless, the last paragraphs of the novel push forward an important ecocritical question: what should a person do with the knowledge he gains about the natural limits of the world? Horacio extracts himself from Ford's project, and in exchange for his silence, he receives passage to Santarem and an undisclosed sum of money. Back in the city, he finds himself changed:

El rumor de la ciudad francamente me molestaba y también despreciaba a esos individuos que no hacían, o no sabían hacer otra cosa, que competir para sacarse el dinero unos a otros y soñar—si lo suyo podía elevarse a la categoría de un sueño—con propósitos vulgares y ridículos (272).

[The hum of the city bothered me, and I detested those people who did nothing, or did not know how to do anything, other than compete for the money of others and dream—if what they did could be raised to that category—of vulgar and ridiculous things (242).]

The capitalist, speculative rhythm of the city bothers Horacio, and he feels not relief but agitation there.

The last snapshot of the protagonist and narrator is provocative. He is plied with offers by foreign speculators, eager to hire him as a labor boss for the next Amazonian enterprise. Horacio must choose among options: cast his lot with the modernizers, go to pre-Franco Spain with Caroline, or find another way. For Latin America's educated elite, Sguiglia insinuates, this is the choice: be complicit with the modernizing mission of outsiders, leave for a struggle against injustice elsewhere, or invent another path. The narration of the novel suggests Horacio's choice: those who know the insider's history of modernization in Latin America must choose a path that includes telling the story of its darker side.

Un viejo que leía novelas de amor

The story Luis Sepúlveda tells in *Un viejo que leía novelas de amor* carries a dedication to Shuar leader Miguel Tzenke and an explanatory foreword about the death of Chico Mendes. Of the three novels about the Amazon region, Sepúlveda's has the most explicit ecological message and the most accurate, individualized portrayal of indigenous cultures. Openly didactic, it depicts the Amazonian world of Ecuador in such a way as to awaken environmental consciousness and to instruct readers in basic facts of tropical ecology.

It is also a meta-ecocritical text. Through the story of the protagonist, the old man who reads love stories, Sepúlveda crafts a conceit about affinities born of reading. By this means, he argues for the importance of literature in awakening a love of nature in a reading public far removed from global hot spots.

Wryly humorous in the tradition of Guatemalan author Augusto Monterroso, the novel reads like a parable of the sometimes complementary, but increasingly disastrous relations between human and nonhuman nature in the Amazon. Sepúlveda's novel has attracted numerous commentaries from scholars interested in environmental literature. Some are excellent literary commentaries, while others use the novel as a springboard for more general discussion on environmental degradation.[11]

Sepúlveda's Amazon is populated by many people, some of whom know and understand it, and others of whom do not. For those individuals who lack understanding, the Amazon appears to rise up to defend itself against violators of its order. For those who know its rhythms, like the aged protagonist Antonio and the Shuar group among whom he lived, the bounty of the Amazon offers a dignified life and death. By contrasting their intersecting lives, Sepúlveda tells the recent history of the Ecuadorian Oriente with an ecological imagination that is ethically compelling.

Sepúlveda sets his novel in a heavily contested portion of the Amazon region. The Oriente is located in the eastern part of Ecuador and abuts the long-disputed border with Peru. Because of altitude variation, the region is incredibly biodiverse. It is also home to several indigenous groups. Some groups are long-standing inhabitants; others are relatively recent transplants from the Ecuadorian highlands. Among those who have a long history in the region are the Shuar, the indigenous group Sepúlveda features prominently in his novel.

The action of the novel centers around the Río Nangaritza, an Amazon tributary that flows through the province of Zamora-Chinchipe in southeastern Ecuador.[12] According to a UNESCO report on a proposed biosphere reserve in the vicinity, "in the area known as Podocarpus National Park and the Condor Cordillera, there exists [sic] more than 7,000 plant species. In addition, only taking into account the Nangaritza River Valley, there exist about one-third of the Ecuadorian Amazon species and 313 bird species"("Proposal").

As was the case elsewhere, the first serious incursions of the capitalist frontier in the Ecuadorian Amazon came with the rubber boom in the late nineteenth century. Rubber incursions peaked between 1890 and 1900, but there was little respite for native peoples and ecosystems in eastern Ecuador (Gerlach 51). Gold prospecting and petroleum exploration followed the rubber boom. Large petroleum reserves were discovered in Sucumbíos province in the Oriente in 1967. Fueled by foreign and national investment, petroleum exploitation ensued.

Around the same time, Latin American regional politics also brought a transformation of the Amazon, as the Ecuadorian government encouraged internal migration to the Oriente to bolster land claims against Peru.[13] The discovery of oil roughly coincided with land reform passed in 1964, and together these developments pushed a stream of migrants from the highlands into eastern Ecuador as *colonos* ["colonists"]. Both the petroleum industry (and the accompanying construction boom of roadways, airstrips, and pipelines) and the promise of land thus opened up the Amazon to new populations from Ecuador and triggered rapid environmental and social change in the Oriente. The Ecuadorian government responded to increasing pressures on Amazonian indigenous peoples by establishing protectorates overseen by outsiders (mainly missionaries). All these dramatically altered the way of life of indigenous peoples (Gerlach 51–52).

Historically, exploitation of oil fields and colonization of Amazonian territory served national political and economic agendas, set by the ruling class of the highlands and coast. The relocation of impoverished people from the highlands into the Amazon alleviated tensions over land distribution. This meant the government did not have to confront large landholders in the highlands in order to quell unrest among the landless. Furthermore, a colonized Amazon region bolstered land claims and acted as a buffer against Peru.[14]

As Ecuador modernized, oil revenue became an increasingly important source of foreign exchange. When oil prices fluctuated in the 1970s and

1980s, first rising dramatically and then falling, the government had to find a way to keep revenues constant and make debt payments. Under pressure from lenders, Ecuador chose to increase the volume of oil exported, and this meant expansion into new oilfields.

As the influx of people competed for resources with established populations in the Ecuadorian Amazon, affected indigenous populations responded with accelerated social and political organization. Indigenous organizations coalesced around two major, interrelated issues: the protection of their cultural identities and the integrity of the ecosystems upon which their communities depended (Gerlach 54). According to Allen Gerlach, "cultural assertion and environmental protection accelerated in tandem" (54). He describes events unleashed in the 1980s and 1990s when the Ecuadorian government granted large portions of the Amazon to oil companies for exploration and exploitation:

> The process of expansion caused deforestation and contamination in a large portion of the Oriente. Little was done to curtail the damage; oil and the land on which it was found were seen primarily as sources of national wealth. Oil companies . . . built roads through the jungle to make the region more accessible to land-hungry colonists (Colonos) from the coast and highlands, and the new arrivals toppled trees to clear fields for cattle and crops. The displacement of the local Indian population continued apace (57).

In the midst of this tumultuous change, Sepúlveda composed and published *Un viejo*.

For readers unfamiliar with the dramas of the Oriente, Sepúlveda's story provides a succinct, fictionalized introduction. The book is quite short, and the plot is deceptively simple. His story conveys the significance of historical actors in determining present realities. Gold prospectors, colonists from the highlands, local and national governments, gringos on expeditions, and the Shuar people, all appear in his narrative.

Like Belgrano Rawson's novel about Tierra del Fuego, *Un viejo* does not contain temporal markings in the form of references to dates and administrations. Instead, it charts the course of the history of the twentieth century in Ecuador's Amazonian territory by looking at the effects of history on an extraordinary place and its people. The absence of such historical markers, along with the characters formed as types, work together to lend to the novel the tone of a parable.

The "viejo" alluded to in the title is protagonist Antonio José Bolívar Proaño. The plot is straightforward; it is the story of a man and a hunt. The novel opens with what appears to be a crime: the appearance of the mutilated body of a blond man, delivered to El Idilio from upriver. The inept, obese, and permanently perspiring mayor accuses the Indians who accompany the body of having killed the victim. Compelled to speak in their defense (the Indians speak little Spanish), Antonio points out that the nature of the wounds reveal the act of a "tigre."[15] He quickly discovers the motive for its revenge found on the corpse: the pelts of several cubs, still of suckling age, stuffed in the tourist's backpack.

When the death toll mounts on the peripheries of El Idilio, the frightened townspeople enlist the services of Antonio to track the cat roaming the edges of the settlement. Blackmailed by the mayor, Antonio finds himself forced to accept the charge to lead the hunt. Sepúlveda uses the hunt to impart lessons of tropical ecology: tree identification, bat habitat, native fauna. He also uses it to draw a contrast between marginalized Antonio's deep, local knowledge of place, and the complete lack of understanding on the part of the corrupt mayor.

Scientifically and ecologically instructive passages reference the rich heritage of local knowledge. Antonio has himself acquired this knowledge from the Shuar, and Sepúlveda uses flashbacks focalized through Antonio to impart lessons about the Amazon. With these passages, Sepúlveda invites readers to follow the transformation of Antonio's understanding of the Amazon, as well as the growth of his psychological attachment to the place.

In designing such a plot, Sepúlveda seizes upon the perfect structure for instruction. Presumably like most readers, the old man is an outsider to the Amazon region he comes to love. When Antonio first arrives as a young *colono* with his wife, he toils mightily with her to cultivate non-native crops on the two-hectare parcel they were granted by the Ecuadorian government:

Se sentían perdidos, en una estéril lucha con la lluvia que en cada arremetida amenazaba con llevarles la choza, con los mosquitos que en cada pausa del aguacero atacaban con ferocidad imparable, adueñán-

dose de todo el cuerpo, picando, succionando, dejando ardientes ron-
chas y larvas bajo la piel, que al poco tiempo buscarían la luz abriendo
heridas supurantes en su camino hacia la libertad verde, con los ani-
males hambrientos que merodeaban en el monte poblándolo de soni-
dos estremecedores que no dejaban conciliar el sueño (43).
[They felt totally forsaken, in barren conflict with the rains that with
every downpour threatened to carry away their hut; preyed on by the
mosquitoes that at every lull in the storm attacked ferociously, swarm-
ing over their bodies, biting, sucking, leaving itchy swellings on the
skin and under it larvae that opened up weeping sores as they made
their way to light and freedom; surrounded by hungry animals that
prowled through the jungle, their bloodcurdling cries making sleep
impossible (32).]

The above passage intentionally repeats the tropes of the Amazonian
"Green Hell," but that image soon gives way to another. The Shuar that save
the colonists from starvation teach them how to use the forest to feed and
heal themselves. When Antonio's beloved wife dies, the grieving Antonio
seeks out the Shuar to live among them.

The Shuar have a deep knowledge of vast tracts of the Amazon because
they move from place to place cultivating forest gardens in cyclical fashion.
Sepúlveda carefully indicates that the Shuar never perceive Antonio as one
of their own (nor does Antonio perceive himself as such). Instead, Sepúlveda
emphasizes that close personal contact with native peoples teaches Antonio
a profound respect for place. This is an important point in the construction
of Sepúlveda's argument for environmental ethics; Western people do not
need to be indigenous, or "go native" for good, in order to model sound
environmental practice. They do, however, need to listen, like Antonio, with
humility and care.

Flashbacks relate Antonio's transformation and the travails he experiences later at the hands of the state when he returns to the village a different person. Sepúlveda crafts a detailed, often humorous narrative so as to
invite identification with Antonio and the human and nonhuman world he
respects. The author repeatedly contrasts wise Antonio and figures like the
mayor (representing corrupt politicians) and oil company tourists (American capitalists) who make confident assertions but know nothing of local life.

The novel mocks opportunistic miners, corrupt mayors, and reckless tourists until the tragic finale, when they are soundly condemned in the voice of the old man.

At the same time he critiques shortsighted interlopers, Sepúlveda vindicates Shuar knowledge systems that become marginalized in a changing political and economic landscape. The contrast in world views has led literary critic Rodrigo Malaver Rodríguez to assert that "la intención narrativa y temática del autor . . . es memoria cultural, re-actualización y re-conocimiento del *Otro*, en este caso de un espacio en peligro, la selva, que un día pierde su armonía por la intrusión del hombre blanco, quien no mide las consecuencias depredadoras de sus actos por el desconocimiento de las leyes que dicta este sabio ambiente, conocido por indígenas que lo habitan y por Antonio Bolívar Proaño, el protagonista" (40) [the author's narrative and thematic intent . . . is cultural memory, the recuperation and recognition of the Other, in this case an endangered space, the jungle, that one day loses its harmony because of the intrusion of white man. He does not take into consideration the devastating consequences of his actions because he does not know the laws that rule in this wise environment, laws that are known by both the indigenous people who live there and by Antonio Bolívar Proaño, the protagonist]. Malaver Rodriguez's interpretation oversimplifies the novel, though, as the critic appears to peg blame exclusively on "el hombre blanco." Sepúlveda's novel portrays much more complex interactions among humans and with nonhuman nature in the Amazon.

The novel portrays a situation in which marginalized migrants from the highlands unleash destruction on lands they do not understand. Some cause harm because they seek a life of self-sufficient dignity like Antonio; others because they want to get rich as quickly as possible, like the gold prospectors. Instead of dismissing these people, Sepúlveda draws attention to the political and economic forces that lead individual actors to negative interactions with nonhuman nature.

Sepúlveda also demonstrates that literary texts can stretch imaginations, educate, and direct readers to desire certain outcomes. He informs readers about Amazonian flora and fauna, the aquaculture and agriculture of the Shuar, and soils that will not sustain intensive agriculture.[16] Sepúlveda also counters myths about the Amazon, for instance, that it is is a timeless, unchanging, primordial landscape. Consider this passage, taken from the portion of the book that describes Antonio's time among the Shuar:

> Viendo pasar el río Nangaritza hubiera podido pensar que el tiempo esquivaba aquel rincón amazónico, pero las aves sabían que poderosas lenguas avanzaban desde occidente hurgando en el cuerpo de la selva. Enormes máquinas abrían caminos y los shuar aumentaron su movilidad. Ya no permanecían los tres años acostumbrados en un mismo lugar, para luego desplazarse y permitir la recuperación de la naturaleza. . . .
>
> Llegaban más colonos, ahora llamados con promesas de desarrollo ganadero y maderero. Con ellos llegaba también el alcohol desprovisto de ritual. . . . Y, sobre todo, aumentaba la peste de los buscadores de oro, individuos sin escrúpulos venidos desde lejos desde todos los confines sin otro norte que una riqueza rápida.
>
> Los shuar se movían hacia el oriente buscando la intimidad de las selvas impenetrables (52–53).
>
> [Watching the Nangaritza flow by, it was easy to think that time had forgotten that corner of Amazonia, but the birds knew that powerful tongues were stretching out from the west, burrowing into the body of the jungle.
>
> Enormous machines were opening up roads, and the Shuar became more mobile. From now on they no longer followed their custom of staying three years in one place before moving on to let Nature recover. . . .
>
> More settlers came, drawn by promises of a future in cattle and timber. They also brought alcohol uncontrolled by ritual. . . . And above all, the plague of gold prospectors grew, unscrupulous individuals who came from every side with the single aim of making a quick fortune.
>
> The Shuar headed east, seeking the seclusion of the impenetrable jungle (42–43).]

The passage begins by acknowledging a facile conclusion a casual observer could make by looking just at the river—that the river is timeless, passed over by history. The perspective then shifts to another element of that landscape, to the birds, the avifauna of the Amazon that could disavow the first statement because they flee approaching change in their migration through the canopy. Then the passage shifts again, this time to the Shuar and their long historical memory of dwelling deeply in place. Finally, Sepúlveda catalogues the degradation, with overtones of sexual violation registered by the Spanish verbs "avanzar" [to advance], "hurgar" [to dig or burrow in], and "abrir" [to open or open up].

Antonio's final condemnation of "the gringo," "the mayor," and the "gold prospectors" echoes the language of sexual exploitation. Here is the last paragraph of the novel:

> Antonio José Bolívar Proaño se quitó la dentadura postiza, la guardó envuelta en el pañuelo y, sin dejar de maldecir al gringo inaugurador de la tragedia, al alcalde, a los buscadores de oro, a todos los que emputecían la virginidad de su amazonía, cortó de un machetazo una gruesa rama, y apoyado en ella se echó a andar en pos de El Idilio, de su choza, y de sus novelas que hablaban del amor con palabras tan hermosas que a veces le hacían olvidar la barbarie humana (136–37).
>
> [Antonio José Bolívar Proaño took out his false teeth, wrapped them in his handkerchief and, cursing the gringo responsible for the tragedy, the mayor, the gold prospectors, all those who whored on his virgin Amazonia, he chopped off a thick branch with his machete. Leaning on it, he set off in the direction of El Idilio, his hut, and his novels that spoke of love in such beautiful words they sometimes made him forget the barbarity of man (131).]

This passage gestures back to the original "crime scene" and reminds readers of Antonio's conservation ethic (he uses his dentures only when he needs them) (Gromides and Vogel). When the thoughtless gringo kills the cubs of the cat and wounds her mate, he unleashes the central conflict and the ensuing tragedies that comprise the novel. Antonio kills the cat and returns alone, successful and sorrowful to the town, to "civilization."

Camilo Gomides and Joe Vogel assert that the climax is crucial for the ecocritical message of the text because it motivates the reader to live within limits of sustainability. I cite here at length their analysis of the ending:

> Antonio and the female ocelot are two trajectories put in collision by forces outside their own control: (1) the ocelot crazed by the brutality of the gringos who have killed its young and (2) Antonio by the bru-

tality of a system that makes him kill what is also an innocent victim, namely, the ocelot. However, the scene is unexpected. The ocelot senses Antonio's "land ethic" and lures him to her mate who is writhing in pain having been shot by one of the tourists. After Antonio performs euthanasia on the male ocelot, the female then turns on Antonio (one daresay with the neoclassical economic logic of marginal benefits/marginal costs). The fact that Antonio survives and not the female ocelot, is didactic. One sees the message by considering the alternative—any victory of the ocelot would imply that the forest will prevail. By having Antonio survive, the urgency of ending an horrific system is put into high relief (13–14).

For Gomides and Vogel, the climax seals the moral implications of the text. Antonio and the cat are wounded in the final battle, and Sepúlveda's description echoes with larger significance: "Estaban iguales. Los dos heridos" (135) [They were quits. Both wounded (129)]. After Antonio kills the ocelot, he weeps, ashamed, and pushes the body of the animal into the waters.

For ecocritical purposes, the ending is effective. As my students observed when I taught this novel, readers do not want either the cat or Antonio to die. And this is Sepúlveda's lasting lesson in ecology: readers should understand themselves as part of an ecosystem and desire its preservation, along with their own.

The last paragraph of the novel also closes the case Sepúlveda has been making for environmental literature throughout the novel. Sepúlveda furthers the moral and didactic ends of the text by signaling literature as an important tool in mitigating environmental tragedy. For this interpretation of the text, the central figure again is Antonio. Antonio's very name is suggestive of a history both of resistance and utopian dreams, a place from which a better future might come. Antonio José Bolívar Proaño carries three names linked to Latin American independence movements: "Antonio" is the first name of the Cuban independence movement leader Antonio Maceo; "José" is the first name of both José Martí, of the Cuban independence movement, and José de San Martín, of the struggle for liberation in southern South America. "Bolívar," is the last name of South American liberator Simón Bolívar. Proaño combines "pro" and "año," a combination suggestive of the future (and also the same last name of a Catholic bishop and priest Leonidas Proaño, famous for his defense of native Ecuadorian peoples).

So, how does Antonio figure in the meta-literary lessons of the text? Antonio venerates the memory of his late wife, and it is her death that teaches him about suffering. It also leads him to his reverence for life and his longing for happy endings. He satisfies this desire by reading.

Antonio goes to enormous lengths to acquire love stories to read, and his favorites are romance novels with infinite suffering but happy endings. In the middle of Ecuadorian Amazon, Antonio's reading of romance novels requires enormous acts of imagination. He can identify with suffering and love, but picturing Venice requires tremendous concentration: "Al leer cerca de ciudades llamadas París, Londres o Ginebra, tenía que realizar un enorme esfuerzo de concentración para imaginárselas" (73) [When he read about cities called Paris, London, Geneva, he had to make an enormous effort of concentration to picture what they were like (63)].

In his conclusion, Sepúlveda offers a meta-ecocritical commentary, that is, his novel addresses the value of art in creating environmental awareness. Led deeper into a world they do not know by an affinity with the protagonist, like Antonio, readers of *Un viejo* must exercise their imaginations to envision plants, animals, and actions. If the story is compelling enough, readers might just emerge with a hunger for happy endings for the present environmental crisis.

Conclusions

Souza, Sguiglia, and Sepúlveda all mock the power brokers of modernization, but only in Sepúlveda's story does the fate of Amazon nature—and not just Latin American politics—appear to hang in the balance. Sepúlveda goes beyond Souza and Sguiglia in a critique of modernity that drives home a moral point about dwelling respectfully in place. At the same time, he argues for the value of fiction in shaping imaginations and sympathies.

Souza and Sguiglia's works express more limited ecological imaginations, and they have ambiguous endings. In their fictional worlds, modernization schemes, led by foreign investors in cahoots with national leaders, all fail in the face of Latin American social and environmental realities. But best practices for the future of human and nonhuman nature are not clear, and neither is the ecocritical message. By contrast, *Un viejo* is a sophisticated ecocritical text that overcomes traditional tensions between ecocentric and anthropocentric visions. And it has a clear conclusion: the love story about an old man and the world he inhabits ends sadly so that readers denied a happy ending might work for a real one.

Chapter 3

Paradise for Sale, or Fictions of Costa Rica

The tropical isthmus of Central America is a region notorious for dictatorship, civil war, and endemic poverty during the twentieth century. Surrounded by more politically troubled neighbors, Costa Rica is a country often presented as the exceptional case in Central America, a land of democratic traditons and breathtaking natural wonder. Indeed, Costa Rica is famous for environmental leadership in the developing world. Politicians and tour operators cite those facts to distinguish Costa Rica from Nicaragua and other nations. Nevertheless, Costa Rica shares much in common with its neighbors, especially when it comes to economic development and environmental transformation.

Costa Rica has an agricultural export economy closely tied to markets in the United States. It boasts great biodiversity, but it also has a high rate of deforestation. The three Costa Rican novels I consider here probe a common, regional, economic and environmental history, and they also explore the advent of the ecotourism industry in which Costa Rica excels. Each of the novels also depicts a struggle over identity and the environment in Costa Rica.

Costa Rican authors bring culture, land use, and politics into the foreground of their stories. They also draw attention to the ways powerful economic forces shape reality in a small Latin American country. These novels include the earliest novel of ecological imagination in the present study, *Murámonos, Federico* (1973) by Joaquín Gutiérrez, as well as *Calypso* (1996) by Tatiana Lobo, and *La loca de Gandoca* (1995) by Anacristina Rossi. All the novels were published in Costa Rica by authors who have enjoyed editorial success and whose works have been influential in literary circles in the country and beyond.

Temporally, the stories span the latter half of the twentieth century, overlapping in time the diegesis of the Amazonian novels and pushing forward to the end of the twentieth century. All turn to a comparatively marginal landscape in Costa Rican national politics, the Atlantic coastal region. In my estimation, the contested Atlantic coast is the vehicle by which authors critique neoliberal economic models and indict the corruption and racism by which they are implemented domestically.

Like the texts about Tierra del Fuego and the Amazon, the Costa Rican novels share an interest in the way Latin America fits into the world economy. *Murámonos, Federico* brings to the fore issues associated with the cultivation of tropical crops for export markets in the United States and Europe. *Calypso* fictionalizes the entire history of the Atlantic coast by taking a small community as a microcosm of change and the process of transculturation that accompanies modernity.[1] *La loca de Gandoca* mounts an ecofeminist critique of the "green" reputation of Costa Rica as an ecotourism destination. Each novel also gives insight into gender politics and the dynamics of power within Costa Rica.

Contextualizing the Novels

In terms of both biodiversity and politics, Central America is unique as the crossroads between two continents. Astounding biodiversity in the form of unique species of flora and fauna fills the narrow isthmus. Extensive coastlines, volcanic mountain ranges, and dramatic variations in elevation and rainfall make for a wide range of ecosystems, from mangroves to tropical dry forests. Geographic position has meant that Central America not only boasts incredible biodiversity but also routinely faces numerous challenges of a geopolitical nature. Further complicating the environmental landscape, endemic poverty means that both individuals and governments frequently pin their hopes on natural resources for personal and national economic survival.

In the geopolitical realm, Central American countries have always felt the influence of more powerful nations, from the era of canal politics to the Cold War.[2] From the earliest European fantasies of an interoceanic route through the isthmus, canal construction captured the imaginations of Central American politicians and local and foreign speculators. As canal politics indicate, the government of the United States hovered importantly over politics and economies in Central America throughout the twentieth century. The fear of communism became a pretext for interventions and invasions that secured U.S. business interests in the isthmus. The most notorious, direct interventions were in Guatemala in 1954 and in Nicaragua, first during the Sandino rebellion of the 1930s and then, during the contra war in the 1980s. The CIA-backed coup in Guatemala in 1954 was deeply connected to the U.S.-based firm the United Fruit Company, one of the largest landowners and monoculture operations in Central America at its height. The twenty-first century brought more democratic governments to Central America, but the economies of the region have remained closely linked to the United States. Trade and, increasingly, remittances from workers living in the United States are the bedrock of Central American economies.

Each economic and political transformation in Central America brought changes to landscape and society. Before European contact, numerous indigenous cultures peopled the isthmus. But by the sixteenth century, the centralized Mayan empire that had dominated the northern regions had long since ceased to exist. When Spanish explorers arrived, they did not find large, urbanized indigenous populations anywhere in Central America as they had in the Andes or Central Mexico. The lack of easily conquered urban centers and mineral wealth meant Central America was a marginal territory for Spain during the colonial period.

The decades after independence brought dramatic change in the form of liberal ideas in economics, education, and politics. Liberal politicians touting civilization and progress stamped their influence on laws and landscapes in Central America in the last decades of the nineteenth century. Liberal leaders encouraged the cultivation of forested lands for cultivation of export crops, and they pushed subsistence farmers into wage labor in a new export-oriented economy. With foreign exchange largely dependent on agricultural exports, land use in Central America changed dramatically. Land and power became concentrated. Monoculture (cochineal, coffee, bananas) and extractive industries (timber) became centerpieces of official economic models. Demand for labor for both clearing land and tending export crops was high, and in some countries, land concentration reached extreme proportions as small farmers and indigenous communities were pushed off their lands into wage labor.[3]

There were important variations in social and economic evolution country by country, but across the board, forest cover diminished, and population growth combined with intensive monoculture to strain the carrying capacity of land and water. Later, the "Green Revolution" that took place in the 1950s and 1960s promised higher yields on less land with chemical-intensive agriculture, but this had deleterious effects on small producers (who could not afford expensive inputs), laborers (who suffered disability and disease from close contact with toxic chemicals), and of course, ecosystems and watersheds.

The 1970s and 1980s were decades of insurrection, repression, and brutal civil wars that decimated Guatemala, El Salvador, and Nicaragua. Through the movement of refugees and illegal and legal arms trade, the wars affected all the countries of the isthmus. In the immediate postwar period in the early and mid-1990s, most Central American governments espoused neoliberal economic models that insisted prosperity and stability would come through free trade. In spite of such promises, throughout the 1990s, economic conditions worsened for the working classes and out-migration increased. Crime rates also climbed, reflecting targeted political intimidation, persistent domestic violence, and growing street crime. Finally, in the new global politics of the war on drugs, the region figured in the crosshairs of competing narcotrafficking organizations and U.S. antinarcotics initiatives.

In the context of turbulent change, environmental issues have often been a low priority for political actors and even many activists in Central America. In contrast to its neighbors, though, Costa Rica quickly built a booming ecotourist industry. Costa Rica capitalized on its democratic reputation in the Central American postwar period. Costa Rica has enjoyed relative peace and stability in the twentieth century, and the 1949 constitution (in effect today) abolished the military. While its neighbors were reconstituting civil society, Costa Rica attracted foreign investment in sectors like manufacturing, tourism, and real estate.

Despite its reputation for relative peace and prosperity in the second half of the twentieth century, Costa Rica shares the same historical dependence on tropical export crops as the rest of Central America. Both the landscape and literature give testament to the environmental consequences of these industries. For example, beginning in the nineteenth century, large tracts of land in the Costa Rican Central Valley (the Meseta Central) were deforested for coffee cultivation, and coastal regions were transformed into banana plantations. Later, when beef exports boomed in the 1960s, deforestation accelerated rapidly as individuals converted lands from forest to pasture: "trees were cleared at over 48,000 hectares a year to make way for a 62% increase in grazing lands" in the decade from 1963 to 1973 (Trejos 71). Reduction and elimination of habitat provoked species declines and extinctions.

Despite these realities, Costa Rica enjoys the reputation in many circles as a success story for environmentalists. Nowhere in Central America is the concentration of biodiversity higher, nor more highly touted, than in Costa Rica. For casual observers in the United States and Europe, Costa Rica is synonymous with rain forests and conservation. Free from standing armies, unscathed by brutal civil wars, and with high literacy rates and a famous tourist economy, Costa Rica looks like a triumph in the effort to create a conservation culture and make it economically viable.[4]

The transformation of Costa Rica into an ecotourism destination has been dramatic. National parks were first established under a minor provision of the 1969 forestry law, thanks to concern among biologists, agronomists, and activists about loss of forest lands and species habitat (Steinberg 58–59).[5] The national parks law (Law #6084) was passed in 1977, and around the same time, Costa Rica ratified several important international conventions related to environmental protection.[6] Its reputation as a green tourist destination grew dramatically in just a couple of decades (Trejos 75). For example, the 1983 country study of Costa Rica published by the U.S. government mentions only bananas and coffee as major industries. There is the briefest reference to tourism and that only in a section discussing government agencies.[7] A few years later, tourism dominated the economy of Costa Rica, and Costa Rica headed the list of ecotourism destinations in countless travel brochures.

How did this happen? Long touted as the "good child" in Central America for its relative peace and prosperity, Costa Rica moved to cash in on its positive image in the 1990s. It sold itself as a peaceful ecological paradise, a welcoming host for multinational companies and a haven for tourists eager to see the wonders of the tropics without confronting the gross social inequities in neighboring countries. Ironically, however, Costa Rica had one of the highest per capita deforestation rates in the world during roughly this same time period (Trejos 91).

Just seven years after the 1983 country study that barely mentioned tourism, much less ecotourism, travel writers were extolling the wonders of Costa Rica:

> And what a place to travel. Visitors can walk among rain forest giants, see green turtles nesting, get a ringside view of one of the most active volcanoes in the world, ogle at the keel-billed toucan, and hear the howler monkey. Pristine beaches beckon on the Caribbean and Pacific. Trees alive with their own miniforests of bromeliads, lichens, and mosses assume mysterious forms in the high cloud forests; orchids grow wild amid lush vegetation that tumbles down among road cuts. Miles of coffee fincas (farms), sugarcane fields, and pineapple and banana plantations bear witness to a rural heritage and the influence of agriculture on the life of the country today (Sheck 2–3).

The glowing review is ripe for deconstruction. Along with the boom in tourism, hotel, resort, and road construction also boomed. It became possible not only to see toucans and the rare quetzal, but also to observe groups of forty or more tourists staring at these creatures in national and private parks.[8] Finally, the writer's comments on the "rural heritage" sound particularly naïve, especially since sugar, coffee, bananas, and pineapple are all ecosystem-altering export commodities mainly cultivated for first-world consumers.

Despite problems, Costa Rica remains a popular "green" destination for casual observers. Costa Rican government officials from both major political parties regularly cite the tourism industry and free-trade agreements as paths to new economic prosperity. However, both new industries share with past economic models a heavy dependence on North American consumption. They leave the country vulnerable to economic problems if there are disruptions in the United States, and their environmental impact is only just now being fully apprehended.

Despite external interruptions and internal dissent, Costa Rica continues to attract international capital, and the country passed the Central American Free Trade Agreement in 2008. Investments by North Americans and Europeans finance assembly plants and petroleum exploration, and they pay for the construction of new tourist centers, housing developments, and mega-resorts. In the 1990s and early twenty-first century, Costa Rican environmentalists and social activists used the Internet to wage campaigns against petroleum exploration on the Atlantic coast, the construction of hydroelectric facililties, the Central American Free Trade Agreement, and the Plan Puebla-Panamá.

Not surprisingly, environmental issues and economic transformation figure as significant concerns in late-twentieth-century novels from Costa Rica. In addition to articulating environmental awareness, the novels I have selected for study also participate in a trend in Costa Rican literature (and Central American postwar literature as a whole) to challenge and deconstruct the image of the nation. Each novel challenges traditional myths of Costa Rican history and identity, such as the notion that small landowners and citizens have provided the basis for a strong, prosperous democracy.

The novels all locate the site of their fictionalized struggle on the Atlantic coast of the country. Each offers a distinct reading of the Atlantic coast, but all emphasize that a dynamic, ethnically diverse society comprises a region that is the scene of "global designs" of interlopers, both those from the Meseta Central of Costa Rica and those from abroad. The authors' fictional representations of the Atlantic coast focus on dynamic human communities that have close, historical connections to the nonhuman natural world. These include English-speaking, Afro-Caribbean descendents of settlers from Bocas del Toro, Panama, and Greytown, Nicaragua, as well as indigenous groups like the Kèköldi people (Palmer 7). By reclaiming marginalized people and landscapes, these Costa Rican authors critique the notion of a homogeneous nation that acquiesces to the demands of a global economy. Their bioregionalist narratives register and amplify voices of dissent.

In particular, the authors show how stereotypes and misrepresentations of the Atlantic coast have been used to facilitate the commodification of the landscape for the enrichment of a small group of beneficiaries. Since the diegesis of the novels spans the latter half of the twentieth century, a time in which the modern Costa Rican state was created and consolidated, the texts also show how national politics and cultural prejudices play into the transformation of landscapes in conjunction with forces of global change. According to Steinberg, "to the casual spectator, Costa Rican politics appear peaceful, predictable, and provincial, belying complex networks of social relations wherein powerful actors may exert subtle but determinative influences over the course of events" (50). According to the novels, social relations influenced by powerful actors shape daily realities on the local and national levels. These affect everything from infrastructure expansion to the enforcement of environmental regulations.

In light of these national complexities, each author depicts cultural and ecological diversity in different ways, though each presents them as interrelated concerns. The authors also employ distinct, rhetorical strategies to defend diversity and criticize the Costa Rican state. Gutiérrez, for example, crafts a nationalist defense of diversity and democracy that Rossi echoes two decades later. Rossi and Lobo craft their stories with plots and language informed by feminism, cultural studies, and postmodern anthropology.

Gutiérrez's text shows the consequences of complex political networks for one small territory and the man who loves it. The novel considers a moment in which banana companies dominated both the economy of Limón Province and the lives of smallholders and independent growers. The novels of Lobo and Rossi, published two decades later, counter popular representations of Costa Rica as a paradise free from social ills and inequality. A reading of the novels together, in the light of ecocriticism, yields important insights into the nature of stories, communities, and cultures in Central America's oldest, most celebrated democracy.

Murámonos, Federico

When it comes to registering Costan Rican language and lore, Joaquín Gutiérrez's *Murámonos, Federico* is an exceptional text. For ecocritical purposes, it is significant because it shows an early appropriation of the rhetoric of nature to critique the complicity of the state with foreign investors in the prehistory of Central American neoliberalism. The language of the narrative captures particular Costa Rican modes of speech in a number of social registers, and it records local understanding of place and nature. Because it weaves an intricate world, the novel shows how economic, social, ecological, and cultural pressures in the twentieth century impact different social sectors.

In Costa Rica, both the novel and novelist are famous. Prior to his death in 2000, Gutiérrez was active as a journalist, novelist, and translator. He was particularly beloved in Costa Rica for his tale for children, "Cocorí." He spent time abroad in Latin America and beyond, and his novel *Te acordás, hermano* won the 1978 Casa de las Américas prize. *Murámonos, Federico* won two national prizes in 1973, the Premio de Novela Editorial Costa Rica and the Premio Nacional de Novela Aquileo Echeverría. Literary critic Sonia Marta Mora attests to the importance of the novel within Costa Rican literary history when she asserts that *Murámonos, Federico* represents "un momento verdaderamente culminante de la novela costarricense contemporánea" [a truly culminating moment for the contemporary Costa Rican novel] ("Joaquín Gutiérrez" 246).

Gutiérrez's novel brings political, social, and environmental concerns to the foreground through a plot in which land-use questions loom large. The story features a bourgeois Costa Rican family of four: father Federico is an attorney, mother Estebanita is the daughter of a Supreme Court justice and is a housewife, and together they have two children, Flor de María and José Enrique. As the narrative unfolds, the family begins to disintegrate. Despite a streak of nonconformity, Federico follows the bourgeois, *machista* norms of aloofness and infidelity, but they cost him his health and his relationships with family members. Estabanita, for her part, adheres to the feminine identity prescribed by *marianismo*; she is the long-suffering wife and mother.[9] In the face of Federico's infidelity, she confines herself to bed and to silence. The children grow up and move away, both physically and emotionally.

Intertwined in this family drama is a tale of North American imperialism, environmental change, culture clashes, and corruption. With innovative narrative techniques and masterful command of local language, Gutiérrez explores and deconstructs cherished Costa Rican identities. He targets the wholesome family unit that serves as the foundational myth of Costa Rican democracy and egalitarianism; he also dismantles the notion of Costa Rica as representative democracy, one that advocates for and answers to citizens. Mora argues that Gutiérrez uses both plot and a discordant narrative to present the sense of a world in crisis: "Esto no se refiere únicamente a su poderosa fuerza desmitificadora de la imagen feliz del campo, del papel salvador de las leyes o del progreso nacional bajo la protección extranjera. Alude concretamente a su estilo particular, totalmente alejado de las jerarquías y convenciones sociales, del decoro y el eufemismo" [This does not just refer to his powerful demythologizing of the happy image of the countryside, the saving role of laws, or national progress under foreign protection. It alludes concretely to his particular style, totally removed from hierarchies and social conventions, from decorum and euphemism] ("Insinuación" 25). Gutiérrez undermines authoritarian discourse by means of a polyphonic narrative that shifts narrative voice and perspective. His recuperation of Costa Rica's multicultural and regional identities unravels the tidy rhetoric of the bourgeois middle class and the state that supposedly represents them. Gutiérrez also articulates a profound sense of place that, in his view, undergirds real citizenship. Bureaucrats complicit with foreign exploitation are alienated from the land, but the troubled protagonist inhabits place deeply and from it, acts in defense of self and society.

Murámonos, Federico is then a story of personal redemption of the protagonist, but redemption occurs in relation to the defense of land and community. In crafting such a redemption story, Gutiérrez lays bare the social and political confluences that have altered Costa Rican landscapes. He simultaneously articulates the resistance of local populations who perceive those same landscapes in ways that lead them into conflict with officials of their own government.

Limón Province is the site of most events in the novel. Part of the Atlantic region of Costa Rica, Limón is geographically and culturally removed from the Meseta Central, home to the capital city of San José. The Meseta Central is a coffee-growing region; wealth and political power are concentrated there, often linked to families whose fortunes come from coffee farming for export. In contrast to the Central Valley, Limón Province is a flat, low-lying area traversed by rivers that tumble down out of the mountains and slow down as the topography flattens out. Rivers flow across fertile plains to a coastline of sandy beaches, tropical forests, and a few mangroves.

Limón Province is also home to the majority of the black population of Costa Rica and is thus culturally distinct from the Meseta Central. Many blacks in Limón identify more with the English language and West Indian culture than with the cultural identity promulgated by Central Valley elites who set administrative, cultural, and educational agendas in the central government. The most famous Afro-Costa Rican author, Quince Duncan, for example, records in his literature a drive for inclusion of Caribbean populations in the story of Costa Rica.[10] The geographical and cultural divisions, as well as politics of racial exclusion that Duncan challenges, all come to bear on debates about the future of the Atlantic region.

A plot of land in contested Limón Province is central to the plot of *Murámonos, Federico*. The land in question is on the Pacuare River in Limón Province, on the Atlantic litoral. The Pacuare River is 108 km long, a mid-sized river for Costa Rica, originating in the Talamanca mountain range, and part of the Caribbean versant (Trejos 46). Now important upstream for the whitewater rafting industry, the river and the land downstream are most significant in Gutiérrez's novel. This is because they shape Federico's identity. For Federico, his chunk of Limón Province is inscribed with memories of his stubborn old grandfather. The plot Federico inherits is also his refuge from the pressures of family and finances, a place where he can see himself as an independent nonconformist and not as a failure according to norms of gender and class.

Early in the plot, protagonist Federico faces the machinations of the "Company," that is, the North American banana company that wants to buy the property. Gutiérrez describes the river land in intimate terms, using an arsenal of metaphors in discourse focalized through Federico. In dialogues between Federico and purchase agents for the banana company, Gutiérrez deploys the powerful metaphors of marriage and home to counter threats to the integrity of the land. These metaphors are the same ones essential to Costa Rican foundational myths of nationhood, often represented as an egalitarian nation of small farmers.[11] Federico is in financial straits, but he rejects the offers of the Company because the commodification of the land repulses him: "Ochenta metros de ancho, sereno, color puma, el Pacuare disimulaba su poderío meciendo juguetón la piragua del Zambo. ¿Es que una maravilla de este tamaño podía tener precio? ¿Reducirse a dólares miserables y porcentajes?" (48) [Eighty meters across, quiet, the color of a puma, the Pacuare hid its power, playfully rocking the Zambo's canoe. Could a marvel of this size have a price? Could it be reduced to miserable dollars and percentages?]. Passages like these juxtapose natural beauty, coupled with human history, to suggest economic terms as meager substitutions for the power of a natural place for human beings.

Just after this portion of the text, Gutiérrez offers an evocative passage about the movement of wildlife near the river. The lyrical passage conveys a close knowledge of flora, fauna, and the rhythms of life. In terms of the structure of the plot, the passage justifies Federico's ardent defense of his land against otherwise compelling financial arguments. Such a structure is significant. Gutiérrez invites the reader, who may or may not share Federico's love of land, to identify with the protagonist's passions and not just his failure and frustrations. More than anything else, Federico's passion resides in a particular place that anchors his sense of self and history. Gutiérrez concludes the section with this image of Federico: "Afirmó las espaldas en un tronco de bambú y se quedó largo rato, bien largo, sin hablar. Sin pensar. El río había cambiado de curso y en esos momentos le pasaba todo su caudal por en medio del pecho" (48) [He leaned his shoulders against a bamboo trunk and stayed there a long time, a long, long time, without speaking. Without thinking. The river had changed its course and at that moment its entire current flowed through his chest]. Federico feels the river, beyond the grasp of financial tabulations, beyond even the domain of language.

Though Federico dabbles in agriculture, cultivating some bananas on a portion of property and keeping a few domesticated animals, Gutiérrez makes clear that the draw of the land is something larger for his protagonist. Federico's history and sense of self are rooted in tracts of forest and the river current. He knows the rhythms of the river, the names and habits of its fauna, and folklore involving nearby families. The geography of his homeland is rich with life and stories, and his sense of self is deeply dependent upon this landscape.

Gutiérrez emphasizes that not all Costa Ricans understand the land in such a way, and he locates the attendant clash of cultures and values at the center of the story. Some in Costa Rica embrace the capitalist dream, while others resist its draw. As Gutiérrez depicts it, the connection with the landscape is the locus of resistance for those who dissent. Consider this passage, from a point in the text in which Federico is already deeply embroiled in a fight to retain his land. Federico relates a visit to his cousin, the president of the country, in which he challenged the president: "¿Cómo vas a entender lo que pasa en el país aquí metido como un cusuco en su cueva? ¿Cómo vas a poder gobernar bien detrás de un escritorio? Venite (120) [How are you going to understand what happens in the country if you are holed up like an armadillo in its cave? How are you going to be able to govern well behind a desk? Come]. Federico's argument is that only those who understand the landscape of the country, all of it, can govern well. Federico specifically points out to his cousin that he needs to appreciate the immensity of the Atlantic coast region, an area traditionally disdained by elites from the Central Valley. Unlike his cousin, Federico rejects the trappings of success common to those of his social class in the Central Valley, favoring a simpler, more independent existence in Limón, where he interacts daily with black and mixed-race residents. For Federico, "El Zafiro," his plot of land, is a retreat and ultimately the vehicle by which he can reclaim his life.

Gutiérrez's character models an alternative mode of existence in the midst of a society beginning to embrace an economy based on commodities and consumerism. When Company attorneys show up to pressure Federico to sell, Federico proposes a trade: "Perdón, perdón—dijo Federico pellizcándose la barbilla—Es que yo no estaba pensando en tierras cuando hablé de cambalache. No estaba pensando en tierras. Lo que se me ocurrió . . . es que tal vez yo podría cambiarle 'El Zafiro' a mister Brooks por su gringa" (42) ["Sorry, sorry," Federico said, scratching his chin. "It's just that I wasn't thinking about land when I mentioned a trade. What occurred to me was . . . maybe I could swap "El Zafiro" with Mr. Brooks for his gringa wife"]. With this dialogue, Gutiérrez exposes and derides the fears foreigners have of supposedly lecherous Latin men, while at the same time he underscores how absurd the Company's proposition is from Federico's point of view. Later in that same encounter, he uses a metaphor from marriage, again in the voice of Federico:

Entiendo que las cosas se venden cuando alguien las pone en venta. . . . ¿qué me diría usted si . . . yo llego un día cualquiera a su casa, toco y le salgo de sopetón con que vengo a comprarle la cama matrimonial? Yo creo . . . que me diría que yo estaba en un error. . . . Pero si de todos modos yo me metiera en su dormitorio y me pusiera a dar saltos encima de la cama para probarle los resortes, qué me diría usted entonces? (43)
[I understand that things are sold when someone puts them up for sale. . . . What would you say if . . . one of these days I show up at your house, knock, and all of a sudden announce that I've come to buy your queen-sized bed? I think . . . you'd tell me I was mistaken. . . . But if I managed to get into your bedroom and started to jump on your bed to check out the springs, what would you tell me then?]

By using metaphors of marriage, Gutiérrez suggests the offer to buy the land is not just an intrusion, but a violation of intimate space.

The defense of the land in *Murámonos* links sovereignty and ecology (or land use, at the very least).[12] Federico remarks in anger that "lo que no debería olvidar nadie, nadie al que todavía le quede un poquito de patriotismo o de decencia, es que estos cuatro potreros alrededor de cinco volcanes se llaman todavía Costa Rica. Todavía no se llaman Costa Rica & Company Incorporated" (45) [what no one should forget, no one with a little bit of patriotism or decency, is that these four pastures around five volcanoes is still called Costa Rica. It's not called Costa Rica & Company, Incorporated, yet]. Selective use of English here (and in all references to the Company) is significant because it overtly marks the collusion between the Costa Rican state and American business interests, with English as the language of business subsuming even the name of the country in the example above.

Eventually in the novel, Federico's country fails him, and this seals the ethical message of the novel. Federico must be betrayed so that he can articulate and signal a path to personal and national redemption. Federico's lawyer informs him that the constitution means nothing against the machinations of the Company, and his own cousin urges him to sell. The message is clear: the Costa Rican state acts in the service of the economic models at odds with the desires of citizens acting from their love of place.

Federico's crises multiply at this point in the text, and his intimate relationships with land and family are threatened at their most basic levels. The land crisis exposes Federico's vulnerability and failures as a father, husband, and businessman. The rhetoric has reached its highest degree, and Federico realizes that both the sanctity of family and citizenship can be violated. He appears powerless as man and citizen to save either family or land.

As crises mount in the story, Gutiérrez makes references to the ecological cost of exploitative land use. For example, on one trip to the capital to organize his financial affairs, Federico carries a bundle of hearts of palm. He encounters an old friend who observes that the *palmitos* are now a rarity:

Se conocían desde los tiempos de las medias cortas y las bicicletas, ya ninguno de los dos era joven, ¡ay!, y los palmitos se estaban acabando. . . . a tantas cosas insólitas que ocurrían, ahora se sumaba aquella lenta y dolorosa tragedia vegetal: la extinción de los palmitos" (121–22).

[They'd known each other since the days of knee socks and bicycles, and now neither was young. Ay, and the hearts of palm were running out . . . on top of so many strange things that were happening, now there was this slow, painful, vegetal tragedy: the extinction of the hearts of palm.]

Dramatic ecological change is measured in a lifetime; the temporal distance from childhood to adulthood is enough to chart the demise of the *palmitos*, trees hacked down for both export and domestic consumption of their tender and delectable hearts.

Firmly located at the center of his identity, Federico's land ultimately plays a crucial role in his recovery. With all legal and financial options exhausted, Federico must sell "El Zafiro," but he is determined to let his land have the final word on the deal. Gutiérrez weaves the story in such a way that the plot to destroy the Company's victory develops slowly. It involves the dreaded "moko," a disease endemic and innocuous in tropical forests but deadly to the monoculture of the banana plantations.

In this denouement, Gutiérrez draws on a long history of disease and vulnerability in the Central American banana industry. Because the story is relatively unknown outside the tropics, it is worth exploring for a moment. Monoculture of tropical export crops leads to enhanced vulnerability to pestilence from bacterial and fungal infections. In Costa Rica, disease has a history of closing vast territories to cultivation and wreaking havoc on company operations. According to Steve Marquardt,

> Airborne and soil-dwelling fungi had thrived in the Central American lowland rain forests long before the coming of the banana industry, and the climatic conditions that fostered them remained the same after the trees were cut down and replaced by bananas. What changed were the vast expanses of banana 'clones,' with identical vulnerabilities far more susceptible to damage by a single infectious agent than complex tropical forests (7).

Historically, the Costa Rican banana industry has been crippled by two major fungal diseases: "sigatoka," an airborne disease, and Panama disease, a soilborne fungus (Marquardt 6).

Panama disease struck first, debilitating Caribbean plantations from the early 1900s until the 1930s and leading to their eventual abandonment (Nelson 144). Companies transferred operations to the Pacific coast, where they had acquired large tracts of land from the government, but the disease eventually struck there as well. Eventually, more than a quarter of United Fruit's farm-labor force in Costa Rica was needed for fungal control measures that included year-round spraying of copper sulfate, a practice that eventually caused soil sterility (Marquardt 7). Finally in the 1960s, the banana companies introduced two new varieties of bananas that were resistant to Panama disease. Their success prompted renewed cultivation on the Caribbean coast and the reemergence of local growers there. Nevertheless, even with changes in patterns of land tenure, "the companies still controlled the system" (Tucker 173). Eventually black sigatoka disease took root, and by the 1980s, production and export of bananas declined as a result. In Marquardt's words, "the sigatoka epidemic was less a natural disaster than a product of industrial-scale, globalized agriculture" (7).

Murámonos charts the casualties of the banana industry and fictionalizes the start of an epidemic outbreak of a different disease: Moko. Moko is a bacterial wilt affecting banana plants that appeared on Costa Rican banana farms in the 1950s (Soluri 201). United Fruit scientists discovered that Moko was spread via plant roots, infected tools (like machetes and knives), and also by flying insects like wasps, bees, and fruit flies (Soluri 202).

In *Murámonos*, the Company mistakenly believes that Limón has natural protection from Moko because of its isolation from the Central Valley. Federico reads that Company officials fear that natural barriers might somehow be breached "por causas ajenas a la decisión y empeño del Ministerio de Agricultura" (130) [by causes foreign to the decisions and efforts of the Ministry of Agriculture]. Federico, who observes his land to the smallest detail, begins to plot his revenge: "Lo he visto. Con estos ojos. Porque lo sé conocer. Porque vive endémico en las especies silvestres en mitad de la selva virgen. Porque en el mismo "Zafiro," en un parchón de la selva virgen, hay" (130) [I've seen it. With these eyes. Because I know how to recognize it. Because it lives endemic in wild species in the midst of virgin forest. Because in the "Zafiro" itself, in a patch of virgin forest, there is some]. Faced with no alternative, Federico sells his land at a much lower price than the Company initially offered. Before vacating the property, he and a friend journey into the forest and fell enough trees for Moko to contaminate the land. The land will be useless for the Company for some time, and later news reports suggest that it might have to be flooded to destroy the Moko.[13] In the concluding pages of the novel, Federico's redemption and the vengeance visited upon the Company come from the land itself. Federico believes the land is resilient, and that it is his best weapon against the bottom line of the Company.

As the novel concludes, Federico makes peace with himself and his family. He walks the streets of the capital with his son. He talks of buying a small plot in the Central Valley. The vindication of Gutiérrez's middle-aged protagonist comes when he channels his inherent rebelliousness and nonconformity against worthy enemies: a greedy foreign corporation and the complicit Costa Rican state.

Through the story of Federico and his land, Gutiérrez charts the local effects of monoculture for export on landscapes, peoples, and histories. In this, the earliest of the novels I have included, we see then a sense and defense of place that will be echoed in later critiques of neoliberalism and globalization. The discourse of nature serves to condemn foreign domination, critique a state subservient to international capital, and warn of growing ecological crisis.

Calypso

By the time Tatiana Lobo published *Calypso*, roughly two decades after Gutiérrez published *Murámonos, Federico*, the fortunes of the banana companies had diminished, the Cold War had ended, and the political landscape had changed for Central America. Rhetoric shifted to topics like open markets and free trade, particularly with the United States and other Central American countries. Migration increased, and tourist and drug economies began to expand in the isthmus.

Published in the midst of the neoliberal transformation of Costa Rica, *Calypso* features antihero Lorenzo as a central figure. Lorenzo's adversary, antithesis, and sometimes collaborator in this story is a collective entity: a resilient coastal community that bears his name. Though it takes shape around Lorenzo's general store, Parima Bay defies and transforms Lorenzo's efforts to control it. In fact, the dynamic between Lorenzo and the town resembles the musical tradition practiced among African slaves to which Lobo alludes in the title. As in calypso, the community articulates a rhythm and story of its own, counterpoint to the wealth, authority, and power Lorenzo wields.

Lobo adds another layer of counterpoint to the narrative by means of her signature style of shifting focalization to challenge the possibility of one authoritative representation of reality. Lobo's discursive innovation and narrative polyphony in *Calypso* is a trait shared by other Central American postwar novels. With the signing of the peace accords in the 1990s, Central American literature changed. Testimonies and political narratives gave way to texts that registered dramatic social consequences related to democratization in an age of globalization. Despite nationalist rhetoric eager to maintain distinctions, Costa Rica faced many of the same challenges as its neighbors in postwar years, and postwar literature voiced a common angst. Without exception, citizens in the isthmus faced the hurdles of political corruption, migration, privatization of state industries, dismantling of social programs, and the growth of drug trafficking and organized crime. Disappointment and bitterness about the shortcomings of democracy and promised economic prosperity ran deep and wide throughout in public opinion and in literature.

Postwar literature from Central America marked social changes with fragmentary narratives about underworlds of vice, crime, and corruption. Instead of presenting the compelling ethical choices of the revolutionary years, authors presented alternative stories of the nation, in which the state is corrupt, defunct, or in disarray. Corruption runs deep in this underworld of criminality, intrigue, and violence. Critic Misha Kokotovic argues that while these postwar novels "do not articulate an alternative to the neoliberal present, they nonetheless constitute a forceful critique of it from the perspective of the foundational figure of neoliberal theory: the sovereign individual" (24). The revolutionary aesthetic of the 1970s and 1980s cedes way to a worldview in which individuals and communities are caught in maelstroms of catastrophic social change and violence. In many of these postwar fictions, cities epitomize the degradation of human beings, their communities, and the environment that coincide with the advent of neoliberal hegemony in Central American states. Lobo's text makes the same critique of modernization via free markets, but it turns its sights away from urban streetscapes to the marginalized landscape of the Atlantic coast.

Calypso begins during the Second World War, in the early dawn of a new era of consumer society in the world. For the Talamanca coast, this is an era in which the region has been "'discovered' by buyers and sellers and preachers and teachers and community organizers and government agencies and land speculators and tourists," in the words of Paula Palmer, author of a folk history of the region (7). The opening paragraph of Lobo's novel links the local world of Talamanca to politics beyond its borders, and it rhetorically establishes a parallel between Adolf Hitler and protagonist Lorenzo Parima. According to the opening passage, "ni Hitler ni Parima conocían la existencia de uno y otro y cada quien iniciaba una invasión a territorios ajenos, a su manera y según sus posibilidades" (11) [neither Hitler nor Parima knew of the existence of the other, and each one began an invasion of foreign territories, each in his own way and according to his own means]. Here the narrator is extra- and heterodiegetic, or outside and above the action of the plot, and links events in the small community to the world beyond. This narrative technique will punctuate the novel. Lobo in this way contextualizes forces that transform the south Atlantic coast of Costa Rica from a nearly autonomous zone of blacks and Indians to a center of beachfront development and a corridor for the international drug trade. Lobo chronicles the give-and-take that happens at each step along the way, questioning not just progress in Parima Bay but the notion of progress itself.

Lobo depicts Atlantic coast communities in a nuanced way that underscores the complexities of social and environmental change. Locals are not passive consumers of new commodities thrust upon them, nor are they merely conscripts or victims of new, dominant economic models. In *Calypso*, marginalized populations are active participants in their own reality: they make choices, change their minds, cooperate at times with authorities and other times, resist them, as it suits them. Lobo's characters are unable to retreat entirely from modernity, and indeed, they even actively embrace parts of it, altering it as they go. Lobo's novel thus guides readers away from a binary vision of development (progress or tradition, resistance or compliance) and into a respect for dynamic communities and the landscapes they destroy, create, and inhabit. Ultimately, the novel suggests that human resilience and natural cycles will be more enduring than the impositions of Lorenzo's modernization.

In the opening pages of *Calypso*, Lobo presents Lorenzo's move from the Central Valley to the coast. This beginning instructs readers about the environmental and social history of Costa Rica. Lorenzo has no land to farm in the Central Valley because his family's holdings have been subdivided into smaller and smaller plots that can no longer sustain the descendants of the original tenants. His story of displacement is a historically accurate portrayal of land use issues in Costa Rica. As Nelson notes,

> By the beginning of the 1900s the amount of virgin public land in the Meseta Central was greatly reduced, and size restrictions were placed on government sales. . . . But wealthy individuals and foreigners continued to buy properties in the decades before World War II, and during this time many small plots were reduced in size by division among heirs to the point where they were too small to support a family (137).

Impoverished and marginalized himself, Lorenzo sets out to make his fortune on the Caribbean coast.

Lobo's antihero is poor, but he possesses a firm belief in capitalism, a shrewd eye for personal advantage, and a potent racism that quells his conscience. Upon arrival on the Atlantic coast, Lorenzo works at Puerto Limón, but he is essentially lazy and resents doing the back-breaking work of loading cargo on the docks. After a short while, Lorenzo and the popular black stevedore Plantintáh (nicknamed for his love of plantain tarts) quit their jobs at the port and start a general store down the coast, using the dowry of Plantintáh's sweetheart as start-up capital.

Lobo scripts the plot opening so that it moves quickly. Lorenzo edges out his black partner and builds his empire in "Parima Bay." But alongside his story of progress and financial success, Lobo presents a parallel discourse in the voice of a sardonic narrator who undermines the validity of Lorenzo's version of events. For example, Lorenzo tells himself and anyone that will listen that Plantintáh only wanted to be his partner to take advantage of Lorenzo's work ethic and intelligence, but the narrator supplies ample evidence to disavow his claims. By drawing Lorenzo's efforts to reshape collective memory into the forefront of the story of his success, Lobo argues that racism, capitalism, and environmental change are fundamentally interconnected. She also unmasks the rhetoric powerful people have used to obfuscate those connections.

Lobo casts Lorenzo as capitalist par excellence, a man too busy attending clients to notice the sea. In fact, he shows disdain for the coastal environment and for all but the Scarlet women, three generations of Afro-Caribbean women with whom he is infatuated. From the beginning, Lobo sets up a contrast between Lorenzo's suspicion of Caribbean cultures and nature and the sense of place and life that organically springs from the people living there. Lobo constructs a world in which Lorenzo's ethics of convenience, driven by greed, constantly clashes with a different force at work in the community, one she depicts as centered on kinship, caring, and dwelling deeply in place.

Lobo establishes this contrast from the beginning of the novel. One of the earliest images of Lorenzo is of a man uncomfortable with the cultural and physical landscape. Lorenzo embarks on a small boat with Plantintáh and other passengers to reach the spot that will become Parima Bay. The boat sets out on a calm sea:

> Los pájaros marinos, pelícanos y gaviotas, acompañaron al velero durante un cierto trecho y luego regresaron a sus rocas, a la seguridad de la playa y a la búsqueda incansable de sus alimentos habituales. Fingiendo una seguridad que estaba lejos de tener, Lorenzo contabilizaba las palmeras visibles a su derecha para no ver la proa levantarse sobre la línea de flotación, hecho que lo ponía nervioso. Un pajarillo descansó sobre la borda, rascándose debajo del ala con su pico (15).
> [The sea birds, pelicans and seagulls, accompanied the sailboat for a while and later returned to their rocks, to the safety of the beach and the insatiable search for their usual food. Faking a confidence he was far from having, Lorenzo counted the palm trees visible to his right so that he wouldn't have to look at the prow rising about the waterline, a fact that made him nervous. A small bird rested on the gunwale, scratching itself under its wing with its beak.]

Here Lobo contrasts a nervous Lorenzo with both his human and nonhuman companions on the journey. Plantintáh and other passengers take in the landscape with camaraderie and ease, and the birds dedicate themselves to the daily search for sustenance. Lorenzo counts, an endeavor rich with the symbolism of linearity and progress. When Plantintáh looks at the bird and predicts rain, Lorenzo scares away the bird to ward off the bad weather of which it is a harbinger. In this deceptively simple sketch, Lobo encapsulates Lorenzo's philosophy: if the natural world indicates a course he does not like, he plows through it or ignores it, refusing to understand its logic and forcing change anyway. In this case, the bad weather comes despite his efforts, and Lorenzo gets seasick. Lobo uses passages like this, which end with the laughter of the passengers, in order to emphasize Lorenzo's status as outsider, a figure at once self-confident and imminently vulnerable to collective and natural forces greater than his will to power.

In *Calypso*, Lobo undermines the success story of the capitalist, neocolonialist entrepreneur, and she does so with biting parody. Lorenzo owns a busy commissary, and eventually rumors subside about his mistreatment of Plantintáh (including Lorenzo's eventual murder of him in an "accidental" shooting) and his obsession with Amanda Scarlet, Plantintáh's beloved. He becomes prosperous, and focalized passages emphasize Lorenzo's disdain for the people and place responsible for his success: "Parima Bay no era la patria de Lorenzo. El nada tenía que ver con la mar ni con los negros" (83) [Parima Bay was not Lorenzo's land. He didn't have anything to do with either the sea or blacks]. Lobo's narrative depicts Lorenzo as quintessential neocolonialist: he measures success in financial tabulations and has no sense of the value of the place he inhabits or of the people who call that place home.

In her rendering of Parima Bay, Lobo crafts a place in which the story of modernity is superimposed upon a community that exists according to other rhythms. A community springs up around the convenience of Lorenzo's general store, and it does not conform to his vision for the landscape:

Algunos ranchitos de maquengue y techos de palma se levantaron entre la espesura, y los terrenos baldíos y selváticos, propiedad de nadie y de todos, fueron delimitados con simbólicas plantas de hojas rojas. . . . Caían los altos sangrillas a golpes de hacha y sus troncos mojados de savia roja permanecían tendidos en el suelo, como cuerpos degollados, hasta que la lluvia, la podredumbre y los insectos hacían el resto. Ahí había demasiada tierra para muy poca gente, reflexionaba Lorenzo, recordando sin nostalgia los modestos potreros de su infancia (29).

[A few palm-thatched houses rose up out of the thickets, and the vacant and forested tracts, property of no one and everyone, were marked off with symbolic plants with red leaves. . . . The tall sangrilla trees fell from ax blows and the trunks, wet with red sap, lay on the ground like decapitated bodies, until the rain, rot, and insects took care of the rest. In this place there was too much land for too few people, Lorenzo reflected, recalling without nostalgia the modest pastures of his child-hood.]

This passage illustrates Lobo's deft manipulation of perspective to create a compelling narrative. She juxtaposes the residents' view of the land (largely utilitarian) with that of Lorenzo (profit-driven), and she offers a third, more skeptical perspective in the voice of the narrator. The knowing narrator com-pares the felled *sangrillas* (*Pterocarpus officinalis*) to bleeding cadavers, while Lorenzo opines that there is too much land for too few people. The passage emphasizes different approaches to land use. The cadaver metaphor empha-sizes the natural costs incurred by local subsistence use, but the actions that really come in for critique are those of Lorenzo, who lives to convert natural "excess" into profit.

The passage is deceptively simple, but embedded within it is an impor-tant documentation of land-use politics in Costa Rica. In Costa Rican envi-ronmental history, mindsets like that of Lorenzo have prevailed over other conceptualizations of land tenure. In the history of nation formation in Latin America, land has been perceived by political leaders as useless unless it rendered tangible economic benefits. Indeed, land rights and title have often been based on not just occupation of territory but on its "improvement," or conversion for economic use: felling trees for sale and for the purpose of extending fields for agriculture. Succinctly elaborated by John Locke in his *Second Treatise of Government*, this conceptualization of property and nature has offered European and mestizo settlers a legal claim to indigenous territories throughout the Americas (19–20).

Lobo's wry sense of humor highlights the quirks of modernization and the changing patterns of consumption that accompany it, a process of trans-culturation about which Néstor García Canclini has written much.[14] Parima Bay residents marvel at banal and spectacular aspects of modern life, picking and choosing parts that suit them. One particularly fascinating change is the arrival of the bouillon cube: "la que más entusiasmaba a Plantintáh era el asunto de los pollos reducidos a su mínima expresión, en forma de cubitos, de los cuales se podia hacer un caldo no tan rico como el verdadero pero semejante" (42) [what most excited Plantintáh was the novelty of chickens reduced to their minimal expression, in the form of cubes from which one could make a broth not quite as rich as the real one, but close]. With the risible example of the bouillon cube, Lobo depicts the process by which local residents adopt articles of convenience, alter their patterns of consumption, and in turn, change lifestyles and land use. In spite of such changes, though, Lobo portrays a local community that maintains a sense of identity not determined by consumer culture. Though they embrace some modern convenience, Parima Bay residents retain a decidedly matriarchal order, foster interdependencies among themselves, express affinity with the sea and land, and cultivate a rich oral tradition about themselves and their place in the world.

Lobo's Latin American community is neither a hapless, passive victim of consumer cultures nor a quaint and timeless enclave. Instead, Lobo's community returns the gaze of outsiders, from entrepreneurs to naturalists, and works the magic of transculturation on their contributions. The novel focuses on the intruders in question and then, like the lens of a retreating camera, it pans out to show the world they visit. Consider, for example, the description of the arrival and reception of an entomologist. The entomologist is a baffling figure that piques the curiosity of the woman who rents him a room, as well as of young Eudora, Plantintáh's daughter. Residents are perplexed that the entomologist refuses to socialize with anyone and instead spends all his hours tracking, capturing, killing, and mounting butterflies. Lobo's scientist is as clueless about his personal safety as he is curious about butterflies. Without the people of Parima Bay, he would have been lost forever to the tropical habitat of his objects of research:

> Sin ojos más que para capullos, larvas, crisálidas y seres pacíficos de lujuriosas alas, el cazador tampoco participaba de las tertulias que los días sábados ocurrían en la cantina de Lorenzo, ni gastaba saliva con nadie. Es cierto que era un tipo inofensivo pero causó tantas molestias que al cabo de muy breve tiempo todos deseaban que se fuera. . . . Causó muchos trabajos y preocupaciones cuando entusiasmado con las bellezas volantes, corría tras ellas, distraído de todo menos las preciosas rarezas, se internaba en lo profundo de la selva y desaparecía (94).
> [With eyes only for cocoons, larvae, chrysalises and peaceful beings with luxurious wings, the hunter didn't participate in the Saturday afternoon chats that took place in Lorenzo's canteen, nor did he waste his

breath with anyone. It's true he was an inoffensive guy, but he caused so many problems that after a very short while, everyone wished he'd go away. . . . He caused endless work and worry when, excited by the flying beauties, he'd run after them, distracted from everything but the precious rarities, he'd plunge into the depths of the forest and disappear.]

In the anecdote of the scientist, Lobo highlights the local community that dutifully organizes patrols to traipse after him, sounding a conch shell to call him back and wishing desperately he would go away. In passages like this, Lobo rewrites centuries of authoritative narratives about natural history by metropolitan scientists and scholars (perhaps even ecocritics). In texts from Humboldt to Darwin, naturalists presented themselves at the center of their explorations and mentioned local people upon whom their research and well-being depended only in the background, if at all. Lobo's text at once rarifies and humanizes metropolitan naturalists and scientists.

Wry humor plays an important role in winning the sympathy of readers who are just as likely to share the scientist's awkwardness in tropical landscapes as they are to share Lorenzo's culture of convenience. In swaying the alliances of readers, the narrative voice again is crucial. The narrator imparts information about the community that Lorenzo does not know (including how they view him), so the reader ultimately understands the community better than Lorenzo. Lobo thus invites the reader to share a sense of complicity with the narrator, who speaks in a voice skeptical of everything, though it is undeniably partial to the local inhabitants. Here again, then, is a contemporary text of ecological imagination in which a narrator or main character exists as cultural mediator, as were the protagonists in *Tierra del fuego* and *Un viejo que leía novelas de amor.*

The following passage is illustrative of the use of humor in building complicity between reader and narrator on the topic of progress. The text concerns Lorenzo's efforts to get a railroad connection extended to the village:

Cuando por fin el dinero estuvo a su disposición . . . para abaratar costos sugirió que los peones no se trajesen de afuera sino que se contratase a los mismos vecinos de Parima Bay . . . Para su sorpresa, nadie se entusiasmó ante la perspectiva de trabajar ocho horas volteando montaña por un salario insignificante, ya que primero había que botar infinidad de grandes y frondosos árboles de madera dura. . . . [El proyecto] murió frente a un enorme jaibillo centenario, de casi dos metros de diámetro, al que ni el golpe de veinte hachas hubieran podido derribar (115).

[When the money was finally at his disposal . . . in order to reduce costs, he suggested that workers not be brought from outside but rather that Parima Bay residents themselves be contracted . . . To his surprise, no one got excited about the prospect of working eight hours blowing up mountains for a piddling wage, especially since first it would be necessary to fell countless large and leafy hardwoods. . . . [The project] died before a centenary jaibillo tree, almost two meters in diameter, that even the blows of twenty axes could not fell.]

At this point, it comes as no surprise to readers that Lorenzo would skimp on costs at the expense of the local population and expect to be thanked for it by them. Nor is it surprising that the community residents refuse to oblige. As readers evaluate statements in accordance with their knowledge of human nature and facts from the text, Lobo introduces an item of environmental significance. She records the casualty of hardwoods, and she documents the obstacles these trees present to the construction of infrastructure. She troubles the narratives of linear "progress" (in this case, literally that of a rail line) and shows the complicated give-and-take, resistance-and-complicity, of land and labor that developers must negotiate as they seek to incorporate new territories into their grand schemes. In Lobo's narrative, technology does not conquer all in an inevitable, straightforward way; instead, it bends to the desires of people and the reality of the nonhuman world. Sometimes it even comes to a halt in the face of a hundred-year-old tree and a reluctant workforce.

Lobo's antihero of progress promotes infrastructure, brokers in convenience items, and throws himself into the import/export business. His avarice prompts changing patterns of consumption and ultimately unleashes unforeseen consequences that drive him from Parima Bay. For example, the already-wealthy Lorenzo eventually exports beef, and so as not to waste space in containers arriving empty, he adds importing to his portfolio: "Para aprovechar el viaje de sus contenedores, inventó la importación de electrodomésticos" (207) [to take advantage of the container space, he came up with the idea of importing appliances]. This import business floods Costa

Rica with electricity-consuming appliances, and it is particularly lucrative for Lorenzo because he can avoid taxes by bribing port officials. Here Lobo draws attention to another important element that lubricates the engine of change: corruption. The import business appeals to Lorenzo not just because he already pays for container space but also because he can externalize costs and maximize profits by subverting government regulations.

In the last portion of the novel, Lobo features two industries driven by consumption preferences in the developed world. Since the late 1980s, the tourist industry and narcotrafficking have both introduced dramatic changes in societies and landscapes in Costa Rica. Lobo fictionalizes the moment of arrival of the tourist industry in the form of a Belgian consortium that wants to buy land where Lorenzo first experienced the culture of the Caribbean coast and in which he felt acutely out of place. Lobo here parodies the kitschy stereotype of the tropical resort:

El belga . . . perdió el habla conmovido por la belleza del lugar. . . . Un islote, situado exactamente frente a la mejor playa, unido a ésta por un invisible sendero de pedrones sobre el cual pasaba el mar jugando a chorritos grandes y chiquitos, le sugirió, de inmediato, un espectacular bar abierto a la luz de la luna en noches de verano, al que se podía acceder convirtiendo la natural calzada en una funcional acera de cemento (212).

[The Belgian . . . stood speechless before the beauty of the place. . . . A small island, situated precisely in front of the best beach and linked to it by an invisible path of rocks over which the sea flowed, washing over it with playful currents, suggested immediately a spectacular bar open under the moonlight on summer nights. This could be reached by converting the natural walkway into a functional cement sidewalk.]

Again, humor comes into play as the Belgian entrepreneur struggles with language and (in the omitted phrases) tries to find adjectives in Spanish to communicate his ecstasy to a distracted Lorenzo. The dynamics of the passage render Lorenzo the more sympathetic figure, as he tries to exorcise the ghosts of the place while the developer babbles in clichéd terms about resort properties.

In the concluding pages of *Calypso*, Lobo depicts the Atlantic coast landscape as increasingly commodified, exploited, and destroyed by modernization. In one of the initial scenes of the novel, Lobo depicts an inaugural trip on Lorenzo's boat, in which passengers celebrate each fish they spot on the journey. By the end of the novel, coconuts, bananas, and cacao have all fallen prey to Lorenzo's avarice. The land is divided and sold for tourist hotels. Barges of hazardous waste burn at sea, and the thick smoke provides cover for the illicit drug trade.

Ultimately, Lorenzo's greed exacts a high price, one borne by the community and himself. Tempted by enormous profits, Lorenzo dabbles in drug trafficking, the ultimate example of high-risk, high-return capitalism. He welcomes the trade in cocaine and profits from it, but it eventually destroys the object of his desire, the last in the line of Scarlet women. Matilda, Lorenzo's daughter for whom he feels incestuous desires, finds a shipping container full of drugs being transported clandestinely by sea. She is killed by the security measures Lorenzo has put in place on the container to protect his contraband.

The conclusion of the novel suggests both a retreat from modernization and the reclamation of place and knowledge in the wake of death and destruction. In her desperation to help Matilda's disconsolate companion, family friend Stella dances, incanting the Latin names of butterflies that she overheard during the entomologist's visit and that she believes summon her ancestors. As she dances,

> Todo alrededor parecía entrar en la danza, mecíanse las palmeras, cayeron algunos cocos, un enjambre de avispas salió de un almendro seguido por un número nunca visto de mariposas azules que sobrevolaron la orilla del mar para perderse nuevamente en la espesura; las cúculas iniciaron un lento movimiento de retroceso tierra adentro, los pájaros alzaron vuelo tras ellas y en desordenada bandada se alejaron de la mar (264).
>
> [Everything around appeared to enter into the dance; the palm trees swayed, a few coconuts fell, and nest of wasps flew from an almond tree followed by a never-before-seen number of blue butterflies that flew over the edge of the sea before losing themselves again in the thickness. The sloths began a slow movement of inland retreat, the birds took flight behind them, and in a disorganized flock, distanced themselves from the sea.]

Stella dances the way she had seen Plantintáh's sister dance in rituals in the forest, and she loses herself in the performance. As she moves, a tidal wave ensues and sweeps away Lorenzo's store and various other structures in the community.[15]

Lobo's conclusion shares with *Murámonos, Federico* the notion of a flood of waters whereby the natural world reasserts power over a landscape humans wish to control. In *Murámonos*, Gutiérrez invented a protagonist that retreated from the modern, bourgeois project embraced by others of his class. Gutiérrez depicts the encroachment of totalizing economic projects in Costa Rica that reach even marginal places like the lower Pacuare River. His protagonist challenges modernization with its own weapons (the law) and then by unleashing the defenses of the natural world (the banana plague). Lobo's conclusion in *Calypso* is different. The community does not retreat from modernity but instead adapts modernity (and is transformed by it) in a process of transculturation. Lobo ends the novel with a snapshot of this process. Stella's artfully performed grief includes the incantation of scientific names to summon her ancestors, and the tsunami ensues and wipes away the centerpiece of Lorenzo's colonization. Even when the store teeming with consumer and convenience items collapses, community life goes on, organically bound together by an identity deeper than the name on Lorenzo's store. In Lobo's conclusion, tragedy persists, but art transforms knowledge. From the synthesis, communities and landscapes are reclaimed.

La loca de Gandoca

Anacristina Rossi's 1996 novel *La loca de Gandoca* features a coastal landscape in peril, not from rising sea waters but rather from encroaching development. Rossi's novel fully explores the social and environmental reorganization of the Atlantic coast region with the advent of tourism. Rossi presents a Costa Rica in which agents of change, motivated by profit, use enduring traditions of racism, misogyny, and domination to advance the capitalist and neocolonialist frontier. Like Lobo and also Gutiérrez, Rossi counters the official representation of Costa Rica in the 1990s as a country administered by leaders that are ethical, democratic, and responsive to constituent concerns. Her novel is particularly significant because it uses ecofeminism to dismantle idyllic depictions of Costa Rican politics and to articulate a defense of the natural world.

Anacristina Rossi is an acclaimed Costa Rican novelist. In 1985, she was the winner of the Premio Nacional de Novela for her first book *María la noche*. *La loca de Gandoca* has sold more than forty thousand copies, an enormous quantity in a region in which the normal run for a book is one to five thousand copies. Rossi has achieved international recognition for her literary production, and an authorized, English-language translation, by Regina Root, is currently in progress for *La loca de Gandoca*.

At its most basic level, *La loca de Gandoca* is the story of a woman who through personal tragedy assumes the role of environmental activist. It is also much more than that. The novel depicts an environmental struggle in the developing world in a country with a mature environmental movement, one that has become institutionalized, for better and for worse. At the time of the novel's publication, Costa Rica was reaping the benefits of a robust ecotourist industry promoted vigorously by President José Figueres Ferrer. According to Steinberg, "the political rhetoric of sustainability had reached such high levels that ordinary Costa Ricans—and more than a few environmental activists— began to roll their eyes in response to Figueres's ongoing environmental discourse, which more often than not appeared designed for foreign consumption" (92). Written in this context, Rossi's story is an exposé of problems in Costa Rica: alcoholism, racism, sexism, and the corruption of elected officials. It is a demythification of Costa Rica as a paradise, "green" or otherwise. Last but not least, Rossi's novel expresses an eloquent, poetic ecofeminism from a Latin American perspective, an important artistic contribution to the chorus of voices expressing environmental concerns in the region.

The plot of *La loca* centers on the Gandoca-Manzanillo Reserve, a protected coastal area in the Talamanca region. Gandoca is located south of the port of Limón, to which it is connected by a coastal road. The area in question in Rossi's novel is officially known as the *Refugio Nacional de Vida Silvestre Gandoca-Manzanillo*, according to the database of Caribbean Marine Protected Areas. This area of tropical lowland wet forest harbors the last intact mangrove swamp on the Atlantic coast and extensive, living coral reefs. It was legally protected in Costa Rica by executive decree in 1985, but as Rossi shows, remains threatened because of corruption and lack of enforcement (Caribbean MPA).

Coastal towns of Talamanca have historically traded with seaside communities in Panama and Nicaragua, as well as with native peoples inhabiting the Talamanca range (Palmer 7). Indigenous groups include the Bribri and Cabecar peoples; the largest group is the Bribri, and they are closely related to their Cabecar neighbors (García Serrano and Del Monte 58). According to anthropologist Anja Nygren, Talamanca was "one of the last places to be conquered in the Americas" (36). Conquest was only completed after the Costa Rican government promoted land colonization and resource extraction in the region (Nygren 36; Rivas 6–7). The Costa Rican government facilitated land colonization by pushing the indigenous inhabitants into more remote areas that came to be designated as the 35-hectare Kèköldi reserve in 1977 (Nygren 36).

Around the time of the publication of Rossi's novel, the Gandoca-Manzanillo Reserve in particular was at the heart of a controversy involving illegal development and protected lands. In an irony worthy of fiction, the development project was headed by the architect of the 1992 Earth Summit in Rio de Janeiro. According to journalist Martha Honey, the wealthy Canadian businessman Maurice Strong and his company Desarrollos Ecológicos were finishing a $35 million resort within the reserve at the same time that the Rio Summit was opening (44). The uproar in Costa Rica centered on Strong's lack of title to the land and the trampling of rights of the indigenous population, since some of the project fell within the Kèköldi reserve and had not been approved by the indigenous association (Honey 44).

Rossi does not make explicit reference to Strong's company, but her story features development plans by foreign investors who plan to operate within the Gandoca-Manzanillo Reserve. Like Lobo and Gutiérrez, Rossi reclaims people and places at the margins of social and political power in Costa Rica. When Rossi writes about the peripheries of her country, she highlights the importance of these regions for their ecological, spiritual, and cultural value, not for their profit potential. Her story draws attention to internal struggles over place and race in Costa Rica, while at the same time it links concerns there with international conversations about place, ethnicity, and environmental justice.

The protagonist of *La loca* is Daniela Zermat, a Costa Rican who comes to Gandoca after a stint abroad. She comes to know the region through the man with whom she falls in love. Both Daniela and her partner seek refuge in Gandoca from a bourgeois society they each reject for different reasons.

In her article on *La loca*, Sofia Kearns argues that the novel is an ecofeminist testimonial that presents "another side" of Costa Rica. Her analysis centers on the story of Daniela's coming-to-consciousness and on the testimonial function of the text. Kearn's analysis is astute; Daniela's calling to speak, to write, and to give witness to destruction figures prominently in the novel, as evidenced in passages like the following: "Ahora no puedo hacer otra cosa más que esperar, abrir bien grande los ojos y ver en detalle cómo extinguen el esplendor del paraíso. Verlo bien y contarlo, para que conste que una vez existió el paraíso" (105) [Now I can do nothing more than wait, open my eyes wide and see in detail how they extinguish the splendor of paradise. To see it well and to tell of it, to testify that paradise once existed]. Expanding on Kearns' analysis, I argue that the recuperation of Costa Rican gendered and ethnic alterity is an important element in the text's critique of neoliberalism and ecotourism. Rossi's picture of ecological struggle in Latin American society is an intimate portrait of Costa Rica, but it is also a multidimensional and complex response to neoliberal rhetoric and global economic policies.

Recognition and recuperation of "otherness" in Costa Rica pervades Rossi's text from epigraph to conclusion. The novel begins with a reference to the Aztec cosmovision in an epigraph that quotes a pre-Columbian poem: "Oye bien, hijita mía, palomita mía; no es lugar de bienestar en la tierra, no hay alegría, no hay felicidad" [Listen well, my daughter, my dove; it is not a place of well-being on earth, there is no joy, there is no happiness]. The quote emphasizes the brevity and sorrow of human life, a frequent theme in Aztec poetry. References to non-Western traditions continue throughout novel, as Rossi identifies the maintenance of natural order with Indian and black residents of the Atlantic coast. Like the authors discussed in previous chapters, Rossi devalorizes the European ("lo europeo") by associating it with ecological and cultural destruction, and she privileges the "indigenous" for a more harmonious and conscientious interaction with the nonhuman world. Such a representation turns Latin American liberal rhetoric on end, since liberal discourses since the nineteenth century have cast native peoples as obstacles to national progress and prosperity.

Though she draws distinctions between Anglo-European, Afro-Caribbean, and indigenous world views, Rossi generally avoids essentialist and romanticized depictions of marginalized peoples in Costa Rica. Instead she presents a range of individuals from indigenous and black cultures. Some advocate development in Gandoca because they stand to gain from it or because they see it as inevitable, and splits in opinion do not necessarily occur along ethnic and cultural lines. By capturing the polyphony of Atlantic coast populations in her text, Rossi deconstructs the romantic notion of a single indigenous or Afro-Caribbean perspective that is timeless, unchanging, and inherently friendly to ecological concerns.

Rossi consistently builds an argument to demonstrate that the inhabitants of Gandoca have long been custodians of ecological integrity found there. Repeated as a leitmotif throughout the text are statements such as the following: "El sitio más hermoso sobre la tierra era de los negros, era de los indios, era Talamanca" (13) [The most beautiful place on earth was that of the blacks, was that of the Indians, was Talamanca]. This simple sentence has interesting implications. First, the use of the imperfect past tense in Spanish communicates a sense of loss. For years Talamanca has been under the guardianship of these groups, and yet the sentence is not enunciated in the present tense. It no longer "is" but rather belongs to a reality already slipping away. Second, in persistently associating Talamanca with its inhabitants, Rossi populates an area constructed in the collective consciousness of the nation as "natural," "empty," and "*despoblada*" [unpopulated], and thus open for development. For example, in a later passage, the protagonist notes that "durante siglos, indios y negros habían mantenido intacto ese litoral" (31) [for centuries, Indians and blacks had kept this coast intact].

The late director of Oxford's Programme in Traditional Resource Rights, Darrell Posey, pointed out that recognition of human manipulation of apparently "natural" areas is important. According to Posey,

> Indigenous peoples' role in conserving biodiversity has been consistently underestimated. In large part, this is due to the failure to appreciate the anthropogenic . . . or humanized nature of apparently pristine or "wild" landscapes. But scientists are increasingly discovering that what they had thought were wild resources and areas are actually the products of coevolutionary relationships between humans and nature (37).

Failure to recognize the role of indigenous peoples in maintaining territories that outsiders perceive as "natural" or "pristine" is a colonialist fallacy.

Outsiders working from the premise that territories exist in a "natural" state have used this fallacy to justify the seizure of lands for resource extraction in some cases and the creation of protected areas in others. Though environmentalists have often been well intentioned, their actions in creating zones of human exclusion do not recognize the agency or history of local peoples. Many local peoples rightfully protest their exclusion from decision-making processes about protected spaces they have long inhabited. In light of this, Rossi's insistence upon the inhabited nature of Gandoca and Talamanca is important because it signals the rights of "Others" to this territory, and it does so in terms meaningful to both local cultures and to Westerners. She depicts the region as a dwelling place infused with history and spiritual significance.[16] In terms of Western notions of property rights, this same land is also their landscape, that is, a place they have labored to sustain and enrich and which they may rightfully claim.

Even as the novel cultivates an association between cultural and biological diversity, it also exposes rifts in the Costa Rican nation. As with the banana trade alliances in Gutiérrez's novel, in Rossi's contemporary, greenwashed Costa Rica, wealthy national elites (usually from the Central Valley) collaborate with foreign investors involved in land speculation and development. Elites work the political system in search of favors that yield personal gain but undermine the rights of fellow citizens, most often those of other social classes. Protagonist Daniela Zermat at one point observes that

> El maestro dice que las comunidades del Refugio son volubles y que el gobierno las ha manejado muy mal, el gobierno sólo se acuerda de que los negros existen cuando quieren sus votos o sus tierras. Con los indios es peor, como ni siquiera votan ni entienden español no tienen que molestarse en mentirles (62).
>
> [The teacher says that the communities of the Refuge are fickle and that the government has managed them poorly, the government only remembers that blacks exist when they want their votes or their land. With the Indians it's worse, since they don't even vote or understand Spanish, the government doesn't even have to bother with lying to them.]

In the case of Talamanca, the advance of modernization comes in the form of the commodification of the landscape via coastal development for the tourist industry. This means the destruction of environment and livelihood for human inhabitants, who are far enough (literally and figuratively) from the center of power (in the capital, the U.S., and Europe) for their voices to be ignored. For those with financial and political power, the region seems remote, but as Aaron Sachs notes, "environmental degradation, even in areas that seem remote, usually carries a high human cost. And . . . that cost is often borne disproportionately by the people least able to cope with it, people already on the margins of society" (6).

Tourism in its traditional forms, Rossi reminds readers, provides few meaningful jobs for local residents. Instead, it pushes them off lands that sustained their communities for generations and into low-wage service jobs that are seasonal in nature. Local communities can craft sustainable ecotourism, but much popular ecotourism only carries the label "eco" and is actually culturally destructive and environmentally unsustainable. Some theorists, like Helen Gilbert, have in fact argued that "contemporary Western ecotourism is based on specific travel modalities that reflect, and even consciously replay, aspects of European imperialism, especially as manifest in the exploration and subsequent domestication of distant natural environments and their native populations" (255). In the case of Rossi's Costa Rica, developers plan to attract tourists to a tropical paradise where they can enjoy the luxuries of home and gaze at the (rapidly disappearing) rich biodiversity, all while being served by the displaced local population.

Throughout the novel, Rossi singles out neoliberal reverence for private property for biting critique. As crusading Daniela is turned away from one government office after another in her quest to save Gandoca, she notes that "la propiedad privada manda" (15) [private property rules]. Neoliberal deference to private foreign investment has much in common with policies that wrought an earlier radical transformation of the region's landscape. In a book about American business in Central America from 1880 to 1930, Lester D. Langley and Thomas Schoonover write that

> Central American Liberals championed material progress. They facilitated the privatization of communal land, advocated policies that hastened the growth of a wage-dependent labor force . . . [and] offered inducements to foreign settlers and financial interests (14).

Neoliberal politicians of the 1990s championed similar policies and aspired to the material progress that free markets would supposedly bring to their countries.

For Rossi, neoliberal policies are successfully imposed on Costa Rica because they draw on legacies of racial and gender discrimination to suppress dissent of local people who live close to the land. Consider, for example, this passage: "No, esta región ya no nos pertenece. Primero dejó de ser de los indios, luego dejó de ser de los negros, después dejó de ser de los costarricenses en general" (99) [No, this region does not belong to us anymore. First it stopped belonging to the Indians, then it stopped belonging to blacks, and finally it stopped belonging to Costa Ricans in general]. Rossi condemns the corruption of elites and appeals to national pride in place in much the same way that Gutiérrez did two decades earlier.

Rossi's most forceful condemnation of corrupt, paternalistic, and racist government officials comes explicitly in paragraphs describing an official Earth Day celebration. Rossi uses the celebration of the "Día del Medio Ambiente" (akin to Earth Day) to reveal the hypocrisy of a government often cited for its progressive environmental record. At an official Earth Day party, Daniela presses a government minister, known for a progressive environmental record, for a response to the destruction of Gandoca. Her persistence eventually elicits a racist remark and personal threat. Here follows a portion of the exchange, with Daniela speaking first:

—. . . Los negros van a pasar de ser propietarios a ser mucamos. Eso no es progreso.
—Mirá, Daniela, aquí entre nos, ¿en qué vas a poner a trabajar a los negros si no es de mucamos? (47)
[—. . . Blacks are going to go from being property owners to servants. That's not progress.
—Look, Daniela, here between us, where are you going to put blacks to work if not as servants?]

Rossi draws the paternalistic and racist undertones of Costa Rican politics into the foreground. Furthermore, the passage explicitly associates environmental exploitation with other forms of oppression, particularly a disregard and devaluation of women and non-European ethnic groups. As Stephanie Lahar notes, "Oppression and repression are sustained by individuals and institutions that are also most often sexist and heterosexist, racist and classist, as well as exploitative of the natural world" (93). Rossi draws sexism into the picture in an exchange in which several acquaintances comment on

Daniela's physical appearance and uncomely breach of etiquette in pushing too far with her questions. In a few brief paragraphs, then, Rossi captures a process by which politically and financially empowered actors objectify women, stereotype Afro-Latin Americans, and commodify landscapes for personal gain.

The passage on the "Día del Medio Ambiente" celebration also highlights the co-optation that can accompany the institutionalization of the environmental movement. As Greta Gaard observed in a published dialogue with Patrick Murphy,

> Institutionalization is a signal that a movement has come to prominence and to a certain level of power, and where there's power, there's always potential for cooptation. . . . Cooptation occurs when the spokespersons for a movement become disconnected from the less visible members of the movement (5).

Rossi's novel asserts that co-optation of the environmental movement has happened on a grand scale in Costa Rica, where government officials and developers brandish a "green" mantra because it sells well to tourists and foreign investors. According to Rossi, not just co-optation but also the historical power structure of the Costa Rican nation widens the gap between local and national efforts for environmental protection. In Rossi's Costa Rica, white officials from the Central Valley exercise political, social, and cultural supremacy over "indios y negros" who recognize and depend upon the biodiversity of the coast for their livelihoods. As Steinberg notes, the "fate of environmental policies is determined in newspaper rooms, governing bodies of national banks, village councils, legislative committees, party headquarters, police stations, agricultural cooperatives, and teachers' unions . . . Individuals with the numerous social contacts have the greatest such resources at their disposal" (16). Because minority groups have more limited social contacts in Costa Rican spheres of national political influence, they have fewer resources to summon in defense of land and livelihood. As a middle-class woman, Daniela has limited social contacts, and the blacks and Indians in the Gandoca region, fewer still. When Daniela takes on developers, she immediately runs into a strong, thick web of relations that reach from the highest levels of government to gossip-mongers in the local community.

Rossi uses both language and literary technique to bring conflict over politics and place to the foreground. For instance, she inverts the centuries-old trope of the labyrinthine jungle that exists in both explorers' chronicles and "novelas de la selva" (novels of the jungle). In *La loca*, the capital, not the forest, is a dangerous maze where doom awaits the protagonist at the slightest misstep. The capital is a space that shields predators and confounds those without "insider" knowledge of the shifting landscape of political and business alliances.

In contrast to San José, the Gandoca-Manzanillo Reserve is a space of beauty, order, purpose, and self-knowledge. Communities that never fit well into the master narrative of national identity, such as Indians, blacks, and persons of mixed race, all call this territory home. Gandoca is also a place hospitable to contemporary refugees like Daniela, a single mother returning from abroad, and Carlos, her partner, a man who has rejected his own privileged upbringing and struggles with alcoholism.

From a feminist perspective, the reversal of tropes is interesting because it dismantles the frequent association in Hispanic literature between upper-class women and closed, interior spaces where they are presumably more secure. Before the advent of feminist writers, Latin American texts often portrayed bourgeois women in enclosed spaces, most commonly the home, walled garden, church, or convent. Beyond such spaces roam only women of questionable morals or lower social status. When bourgeois women transgress the enclosed space, they are menaced by real danger, social stigma, and their own fears. In Rossi's ecofeminist novel, however, the active, middle-class female protagonist of *La loca* retreats to the natural world where she senses the protection of Yemanyá.[17] Daniela feels threatened indoors in government buildings, offices, hallways, and restaurants. The cities, buildings, and offices of "civilization" are coded in Rossi's novel as spaces of deception and plotting. In them, brokers and beneficiaries of neoliberal policies regulate, control, and exploit the human and nonhuman world.

While she mobilizes nationalist and populist sentiment in defense of the environment, Rossi also knits together an elegiac, feminist ecopoetics in descriptive passages. She intersperses lyrical passages on the beauty of Gandoca with indictments of ecological destruction, making the former even more poignant and the latter more urgent. The celebratory discourse counters centuries of encoding in Latin American literature in which writers represented natural places as blank, empty spaces in need of domination (the "Green Hell" metaphor) or as sites of abundance ripe for exploitation (the "Edenic" or "El Dorado" metaphor). It also responds to decades of travel writing, particularly by naturalists, who saw the world of Latin America as a space to be catalogued and thereby discursively controlled or as a basis for philosophical reflections on the nature of civilization and progress. For example, when Mary Louise Pratt studied the travel writing of nineteenth-century "emissaries" of European science and progress, among them Alexander von Humboldt, she described "the eye scanning prospects in the spatial sense" and knowing itself "to be looking at prospects in the temporal sense" (61). Pratt has held that these "visual descriptions presuppose—naturalize—a transformative project embodied in the Europeans" (61). Through the discursive acts of European naturalists, Latin American landscapes became pristine spaces to be exploited or conserved, and human beings, one more object in the gaze.

In contrast, Rossi's protagonist dwells in place. She does not simply gaze; she hears, smells, feels, and belongs to the world about which she writes. Daniela is the vehicle by which Rossi recodes the natural world in a textual and linguistic economy divorced from market values and the possessor's gaze. Numerous passages in the novel de-privilege the visual, locating it among the other sensory mechanisms by which humans perceive the world. For instance, Daniela queries,

> ¿Qué es la vida silvestre? ¿La arena dorada y las plantas fósiles son vida silvestre? ¿Es vida silvestre el mareante olor de los lirios salvajes? ¿El silencio de las cinco, previo al concierto de pájaros, también? ¿Y el silencio nocturno? (16)
> [What is wild life? Are golden sands and fossilized plants forest life? Is wild life the intoxicating smell of the native lilies? And is it the silence at five, before the concert of the birds? And night's silences?]

In passages such as these, Rossi articulates what Greta Gaard, following Val Plumwood (138–39) describes as "the ideal relation between humans and nature," that is "a relation that requires a kind of speaking with and listening to nature; a relation of human embeddedness in nature that is physical, psychological, and spiritual, and which still preserves the distinct identities of humans and nature" (Gaard and Murphy 3).

This same language echoes through contemporary communiqués issued by indigenous groups in Costa Rica in response to projects from hydroelectric plants to oil pipelines. Consider, for example, the words of the "Manifiesto de las comunidades indígenas afectadas por el eventual Proyecto Hidroeléctrico Boruca" in 2001: "nuestra historia, nuestra identidad y nuestra cosmovisión están íntimamente ligadas a la tierra, los ríos y a toda manifestación de la naturaleza en nuestros territorios (la separación de nuestros territorios significaría para nosotros la muerte, el fin de nuestra historia) (22) [our history, our identity, and our worldview are intimately linked to the earth, rivers, and all natural life in our territories (separation from our territories would mean for us death, the end of our history)]. Rossi's novel constructs a female protagonist acting from a similar awareness of alterity, based on gender and a deep connectedness to place. From this subaltern position, her protagonist challenges the logic of modernization by means of neoliberal reform and integration into a global economy.

At the conclusion of the novel, the defeated protagonist is mired in grief at the loss of her partner and the desecration of Gandoca. A community member urges her to write. In a move that brings readers full circle to the beginning of the novel, Rossi's last page has Daniela write the sentence with which the book opened. In this manner, Rossi widens the circle of interlocutors that must appreciate and protect Gandoca, and her closing paragraph is a call to consciousness for environmental justice.[18]

Conclusions

Gutiérrez, Lobo, and Rossi all draw from communities on the margins of the modern state to give an alternative representation of contemporary Costa Rica. The authors employ rich cultural traditions in Central America in their stories, and in so doing, they reveal the existence of injustices and inequities that are perpetuated by modernization without true democratization. In this way, they introduce readers to a new Central American literature with a richly nuanced perspective on struggles for social and environmental justice for its communities.

Chapter 4

Futuristic Narratives and the Crisis of Place

Crisis of place is the last chapter in the story of modernity, according to the futuristic narratives included in this chapter. The Costa Rican authors of Chapter 3 portrayed neoliberal economic models as advancing menaces to society, but the authors of futuristic tales peer into the future and predict a society these same models will have driven into an ecoapocalypse in the twenty-first century.

¿En quién piensas cuando haces el amor? (1996) by Mexican author Homero Aridjis and *Waslala* (1996) by Nicaraguan writer Gioconda Belli are both futuristic narratives grounded in troubles presently felt or feared in Latin America. The ecological imaginations of Aridjis and Belli take readers to the brink of an apocalypse precipitated by the inexorable logic of markets and the definitive chasm between wealthy and poor in the world. At the same time, each author plumbs ancient and recent Latin American history for alternative models of being and regeneration.

Aridjis and Belli take present-day realities and project them forward in time and scope to argue that the ultimate, logical outcome of modernization is a failed society in a ravaged landscape. Aridjis offers an urban narrative set in a fictionalized Mexico City and anchored in Mexican mythology. For her part, Belli locates her story in the rural tropics, in a fictionalized Nicaragua. Poetic and revolutionary traditions inspire her representation of crisis and restoration.

In these narratives, Aridjis and Belli envision a crisis of self and place.

Human societies approach a point of no return in a world in the throes of environmental crisis. Both authors link societal well-being to the health of ecosystems. Their critique of modernity features sick societies locked in exploitatiave relationships to the natural world. Boredom, alienation, and violence proliferate among people out of touch with the ecological sublime. Interestingly, though, both novels conclude with regenerative gestures inspired by Latin American cultural history. According to this vision, geological, biological, and poetic rhythms undergird life on earth, and an innate biophilia, or natural affinity for life, in human beings offers hope for the renewal of society.[1]

Contextualizing the Novels

Aridjis and Belli profile two different paths of modernization in Latin America; both produce social and environmental crisis. Aridjis portrays the failure of the industrialization and development model in *¿En quién piensas cuando haces el amor?* In *Waslala*, Belli depicts exploitation of Latin America via neocolonialism. Historically, neither model has been conducive to a robust environmental agenda. Indeed, each intensively extracts resources and generates gross inequality, both locally in Latin America and also on a global scale.

Neocolonial economic and social relationships determine reality in Faguas, the fictionalized Nicaragua of Belli's novel. According to Indian environmentalist Ramachandra Guha, "the age of empire had been governed by the belief that white was superior to brown or black," and therefore, certain peoples were meant to enjoy material gain and comfort (*Environmentalism* 65). That belief shapes the neocolonial order in Faguas, too. There, ordinary people struggle to survive fratricidal wars while local strongmen prosper in collusion with wealthy world powers. Faguas supplies oxygen and recreational drugs; it imports out-of-date weaponry and discarded consumer goods.

In contrast, in Aridjis' Moctezuma City, the development dream has run amuck, and the people and environment languish under the toxic skies of the capital. Promoted after the Second World War, during an age of decolonization in Africa, Asia, and the Middle East, the development model argued that all nations could attain levels of affluence enjoyed in the developed world (Guha 65). Optimistic proponents argued that science and technology would lead societies out from poverty (Guha 65). Political leaders championed development and mobilized legal, financial, technical, and human resources of the nation-state in pursuit of the dream. Unfortunately, in most of Latin America, corruption and poverty persisted, national debt grew, and the environment became more toxic.

In taking on the failures of modernization and the oppression of neocolonialism, Aridjis and Belli call for environmental justice at a global level. According to Robin Morris Collin,

> Environmental justice challenges the full spectrum of disproportionate impacts which place a toxic boundary around communities of color and vulnerable individuals, making them acceptable sacrifice zones. . . . Many of the same issues raised domestically [in the United States] by the environmental justice movement are posed even more starkly at the international level. . . . Ultimately, the issue of environmental justice is one of global equity (7).

Latin America has been a place of profound inequity at least since the era of colonization, but the crisis deepened during the neoliberal era during which Aridjis and Belli penned their novels. National political and economic elites have drawn toxic boundaries around their own citizens, enriching themselves in collusion with exploitative and polluting industries like mining operations and metal refineries. Sacrifice zones now cover huge swaths of territory and entire cultures: hydroelectric projects displace indigenous peoples, gold mines replace rural communities, and new trade agreements force small farmers off their lands and into a growing stream of migrants.

Indeed, by the mid-1990s, neoliberal economic policies and globalization exacerbated an already wide gap between the wealthy and poor in Latin America. Beginning in the 1980s real wages in Latin America decreased dramatically in nearly all countries. According to economists and historians Jeffrey Bortz and Marcos Aguila, "globalization and the politics of export-led growth accompanied the downward flexibility of wages, the increasing inequality of incomes, and growing wage differentials" (131). Economic inequities grew, and displaced workers moved to urban centers and crossed national boundaries (from Mexico into the United States, from Nicaragua into Costa Rica).

At the same time, environmental degradation accelerated in many parts of Latin America. Increasing inequality of incomes in Latin America may appear irrelevant to environmental debates, but in fact ecological economists argue that this information is quite germane. According to researchers like James K. Boyce, economic disparities lead to an exacerbation of environmental ills. Boyce uses a political-economy framework to analyze environmental degradation and holds that "if the winners are relatively powerful, and the losers relatively powerless, more environmental degradation will occur than in the reverse situation" (2). Belli and Aridjis put their ecological imagination to work to critique a world of powerful winners and increasingly disempowered losers. They imagine a world in which inequities reach such gross extremes that apocalypse nearly ensues.

Inequity is readily apparent in the Latin American metropolis.[2] Millions in the Americas make their home in megacities, where elites occupy historic centers and posh new developments, while poor residents fill concentric rings and hillsides around the city center. In the 1990s and into the twenty-first century, cities have been sites of growing inequity and mounting violence. According to one study of six major metropolitan areas (Buenos Aires, Lima, Mexico City, Montevideo, Rio de Janeiro, and Santiago), "the welfare of most people in the six cities in terms of poverty, income inequality, and unemployment showed little improvement in the 1990s" (Roberts and Portes 59). During the same period, crime and violence grew worse, and "the wealthy increasingly segregated themselves in gated communities and through the use of private health and educational services" (Roberts and Portes 60). Other studies confirm connections between inequity, urbanization, and the uneven distribution of environmental protection measures. According to one research team, "the urbanisation of the so-called 'developing countries' leads to two apparently contradictory trends. On the one hand, there is the introduction of ever more sophisticated technical and institutional environmental protection mechanisms. On the other hand, it is obvious that environmental protection operates in parallel with the widening gap between the rich and the poor. These observations are valid at the national level, between rich and poor countries, and within each society and its social stratas" (Bolay et al. 628).

The segregated, polluted world of the megacity is the reality that confronts millions of new migrants each year in Latin America. Droves come to cities when increasingly desperate conditions force them to leave rural areas in search of income from industrial or service work in the cities. Moderniza-tion programs and trade agreements like the North American Free Trade Agreement (NAFTA) favor intensive agricultural cultivation for export, and this export-oriented model displaces subsistence farming and smaller-scale farms that produce for household consumption and local markets.

Rural-urban migration patterns in Latin America have produced enormous cities plagued by hazards to individual and community health. Researchers point out that the "issues of poverty, overcrowding, poor sanitation, contam-inated water and overpopulation are acutely felt in urban settings" (Iyengar and Nair 339). A *New York Times* article in March 1992, for example, noted that in Mexico City the government "closed the schools . . . , ordered indus-tries to cut their operations sharply and banned almost half of all cars from the streets" (Golden). Just three years later, World Bank researchers noted with concern that "Mexico City is the most populous and the most polluted metropolis in the developing world" (Faiz et al. 303).

In this context, Aridjis' dystopian novel functions as parody and critique of ecocide via modernization. Dystopian narratives represent life in deplor-able conditions of disease, pollution, or poverty; they exist in contrast to utopian novels, which present an ideal society. According to critic Miguel López, the dystopian novel experienced a resurgence in various parts of the world in the late twentieth century. López argues that

la forma de la novela distópica, por su crítica inherente al uso indis-criminado de la tecnología y de la industrialización sin regulación, se presenta como un marco propicio para cuestionar el valor de la global-ización neoliberal y sus repercusiones en el ámbito político, cultural y ecológico (175).

[by means of its inherent critique of the indiscriminate use of technol-ogy and of industrialization without regulation, the form of the dysto-pian novel is an appropriate means by which to question the value of neoliberal globalization and its repercussions in political, cultural, and ecological spheres.]

As noted by social scientists, modernization "introduces an all-enveloping dynamic that disrupts the human and material landscape by imposing new political imperative (market liberalisation, for example) [sic], and sets new objectives, such as increased international competition, which in turn generates new forms of relations between cities, social groups and individuals" (Bolay et al. 629). López also points out an important distinction between Mexican dystopian works and those from the European tradition. He argues that Mexican novels tend to represent not inevitable apocalypse but rather a viable alternative that arises for humanity from the past (174).

Aridjis and Belli fit this pattern in that they lead their readers into an apocalyptic world, and then out again. To make their critique of neoliberal models, Aridjis and Belli use arguments that presuppose a connection between ecological integrity and biopsychosocial well-being in humans. For Collin, "a vision of a future that is not an apocalyptic end to all life requires understanding our legacy—our inheritance in the West as the privileged children of a brutal past—and understanding the claims of environmental justice" (3). Belli confronts a brutal past, but she also celebrates the dreams of those relegated to poverty by an unjust world order. Aridjis condemns the apathy of humans and the self-absorption of politicians who cannot or will not read the signs of the ecological apocalypse all around them. And despite the dystopias they present, the endings of each novel suggest the possibilities of regeneration anchored in the rich history of Mesoamerican societies.

¿En quién piensas cuando haces el amor?

A prolific and acclaimed contemporary writer, Homero Aridjis is also an environmental activist, perhaps the most widely known in all Latin America. Twice president of PEN International, he was a founding member of the Grupo de los Cien, an organization of writers, artists, and scientists that has advocated for environmental concerns in Mexico. The Grupo de los Cien made its first public appearance in Mexico in 1985 with a manifesto calling for action to combat pollution. Among the prominent signatories were artists like painter Rufino Tamayo and Nobel Prize–winning authors Gabriel García Márquez and Octavio Paz (Russell 66–67).

Led by Aridjis and his wife Betty, the Grupo de los Cien sounded the alarm on the toxic environment in Mexico City and advocated for concrete environmental reforms. The group successfully pushed for government protection of three migratory species for which Mexican territory is important: the monarch butterfly, the sea turtle, and the gray whale. The Grupo brought together a coalition of environmentally minded researchers and activists in

1991 and again in 1994. These meetings produced the First and Second Declarations of Morelia, both of which brought international and national publicity to environmental topics of concern in Latin America (Russell 68).

Apart from the Grupo, Aridjis has used his weekly column in the newspaper *La Reforma* to call attention to environmental travesties in Mexico and to condemn political corruption and private greed behind the degradation. In his poetry, Aridjis has long celebrated the vibrancy of the natural world and lately borne witness to the demise of beloved landscapes and spaces. One imperiled place important to Aridjis is the monarch butterfly habitat of his native Michoacán; another is the capital city where he has made a home for many years. According to poetry critic Niall Binns, "apocalyptic warnings and nightmare images of Mexico in the future abound in Aridjis' most recent books. . . . Language is contaminated. Human beings are spiritually crippled, unable to love" (131). What Binns observes for Aridjis' poetry is also true for his narrative, particularly the apocalyptic novel I consider here.

In *¿En quién piensas cuando haces el amor?*, the nonhuman world is barely visible at all, and humans as spiritual, creative beings have almost disappeared along with it. The biological world has nearly been obliterated by technological advance, unchecked industrialization, and pollution without regulation or limit. With the extinction of species, there is an accompanying, related extinction of cultural life and diversity.

Aridjis' choice of themes resonates with the ideas of social ecologist Stephen R. Kellert. Kellert posits that with species extinction, "far more may be at stake than just the dimunition of people's material options. The degradation of life on earth might also signify the possibility of diminished emotional and intellectual well-being and capacity" (7). Aridjis' novel echoes Kellert's point; all but a handful of humans in Moctezuma City are automatons, caught in the cogs of a system too big to imagine or confront. Housed in identical buildings in indistinguishable neighborhoods, city residents live a miserable existence regardless of social class. They inhabit a world where food is unhealthy and unnatural, politicians are corrupt, and police brutalize longtime residents and recently arrived migrants alike. The crime-plagued, sinking city is a magnet for sex tourists and desperate migrants from outlying areas.

In Moctezuma City, ecocide is nearly complete. Trees have disappeared, and only a handful of elements of nature are visible: passing butterflies at a funeral, a garden in dire need of never-delivered seedlings, a last haven of birds and plants kept by three sisters who are friends of the protagonist Yo. The only natural reality that the residents seem to notice is a geological one: the persistent tremors and threats of major quakes. Despite telluric reminders of place that shake the literal bedrock of their world, city dwellers have lost all sense of belonging. They spend their free time pushing through crowded thoroughfares to identical apartments, where they spend hours watching the "Circe de la Comunicación," a never-ending stream of interactive entertainment devoid of content and meaning. Powerless and bored, city residents submit to the basest of human desires. Most consume a steady diet of bad food, casual sex, meaningless spectacle, and random violence.

The novel opens with Yo chomping on an apple as she and her friends receive news of the death of one of the sisters, Rosalba. The apple-eating act cloaks Yo in the symbolism of Eve in Eden and prefigures her importance in the apocalypse-rebirth in the conclusion of the novel. The pages between beginning and end write the urban landscape through Yo's filter as she walks through the city to reach Rosalba's house.

In the midst of ecocide, only the protagonist Yo and her theater-crowd of friends preserve a memory of a lost connection to the natural world, as well as hope for its resurgence. Yo embodies both modern alienation and the possibility of community and consciousness of the persistent rhythms of the natural world. This is because the protagonist literally does not fit in; she is too tall. From her physical vantage point, as well as from the interpretive lens of her memory, she sees the reality of the city.

Apart from her height, the source of Yo's difference is her late father, a man who died clutching the obituary of the last Bengal tiger. During her adolescence, Yo's father Ariel instilled in her an appreciation for both nonhuman nature and the rich history of human beings.[3] As Yo walks through an urban dystopia after his death, she conserves Ariel's affinity for wildlife and his memory of a nearly forgotten past in Moctezuma City. For example, Yo explains that

> si mis pies no habían recorrido en su integridad el planeta que era Ciudad Moctezuma, en mi fantasía muchas veces había llegado hasta el final del Paseo de la Malinche, en cuya glorieta de plantas marchitas y sin chorros de agua, según mi padre Ariel, quien había leído a fray

Diego Durán, cronista de los antiguos mexicanos, estaba el oratorio de Toci, la diosa de los temblores, en cuyo pecho palpitaba el corazón de la tierra (113).

[if my feet had not covered in its entirety the planet that was Moctezuma City, in my imagination I'd arrived many times to the end of the Paseo de la Malinche, where in the middle of the traffic circle filled with wilted plants and missing its fountain, according to my father Ariel, was the oratory of Toci, the goddess of earthquakes, in whose breast beat the heart of the earth.]

Here in Ariel's memory is conserved the notion of Mexico City as the center of the creation, the umbilical site of world.

Despite Ariel's efforts to anchor Yo in time and place, after his death, Yo wanders aimlessly until she is employed in the theater. There she finally finds a job, an identity, and a calling. She is a light technician, illuminating the actions of others: "Donde pones el ojo pones la luz y así todo el mundo verá lo que tú ves—me dije" (66) [wherever you look, you will shed light, and so all the world will see what you see, I told myself]. The passage illuminates precisely the function Aridjis has for the narrator. Her vision of Moctezuma City shines light on environmental degradation, social alienation, and a tenacious hope for resurgence.

One of the chief ecocritical contributions of the novel is its representation of complex webs of culpability for ecological degradation. Aridjis does not condemn technology so much as he denounces corrupt power brokers and lackadaisical citizens. In this brave new world of 2027, estrangement results when people lose their connection to one another, their history, and their place. Technology can be an alienating force and it can be a substitute for real pleasure, but responsibility for the catastrophe in Moctezuma City lies with many.

One agent of catastrophe is the mayor of Moctezuma City, a figure so disconnected from reality that he is a risible caricature of the lying politician. Drawn straight from the pages of the best political satire, of Mexico or anywhere, is this appearance: "Hacia las cuatro de la tarde, el alcalde Agustín Ek y sus funcionarios, con trajes de seda, chamarras negras y zapatos lustrados, emergieron de la Plaza del Cacique Gordo" (28) [Around four o'clock in the afternoon, Mayor Agustín Ek and his officials, dressed in silk suits, black coats, and polished shoes, emerged in the Plaza of the Fat Chief]. As they cross a pedestrian bridge, Ek and his lackeys hold their noses and look down upon sewers, crowded streets, stray dogs, and a decrepit Baroque church. Then Ek declares: "Nuestra ciudad es fértil, no cabe duda. Hemos sembrado progreso en los campos, no cabe duda. El progreso tiene cien metros de alto, no cabe duda" (29) [Our city is fertile, without a doubt. We have sown progress in the fields, without a doubt. Progress is one hundred meters high, without a doubt]. The ironic use of rhetoric that draws from an agrarian past—fertility, sowing, fields—is striking, as is the repetition of the refrain "without a doubt." Aridjis signals here the extent to which the biological world becomes present only as metaphor in a technological society. He echoes here a concern of Buell who notes that "how we image a thing, true or false, affects our conduct toward it" (*Environmental Imagination* 3). When progress (precisely one of Buell's examples) is associated not just with "transit" but also improvement, it affirms associations inherited from politics and industrialization (3). The extreme outcome of such "progress" in Moctezuma City, of course, is the "fertile" city and "improved" countryside.

Other people are unwitting participants in the world of social and ecological degradation. In passages that give snapshots of children and indigenous women who cross Yo's path, Aridjis acknowledges many in Latin America who are too disenfranchised to do anything about the world in which they struggle just to live. At the same time that Aridjis humanizes their figures, he suggests that the true authors of environmental problems are the political and economic elite. Experts in sustainable development point out that "when households lack the means to protect natural resources, capacity factors (poverty) are at fault. When households lack the desire to protect natural resources, incentives are at fault" (Swinton, Escobar, and Reardon 1869). Capacity factors are to blame for the environmental consequences of Aridjis' impoverished masses, while the middle class, mesmerized by digital media, lacks incentives to protect the natural world. Indeed, Aridjis sketches a city composed of the perfect population for ecological apocalypse: a corrupt elite concerned only with maintaining power; a class of passive consumers sated by banal entertainment and food; and a substrate of the desperately poor and dislocated.

To depict and condemn the megalopolis, Aridjis draws heavily from mythology, both Greek and pre-Columbian, and also from theater of Golden Age Spain. The "Circe de la Comunicación," for example, is an allusion to *The Odyssey*. A caricature of digital entertainment, the Circe turns viewers into "puercos mentales" [mental pigs], ensnaring them with its spectacle and rendering them powerless, just as the mythological Circe did. "Circe" is also a play on words in Spanish: "asirse" means to seize or grab hold of something or someone. The Circe is one element of the novel that has piqued the interest of critic Hélène de Fays. For de Fays, Aridjis' depiction of the Circe is indicative of neo-Luddism in the novel. She argues that Aridjis "portrays Moctezuma City, a metaphor for Mexico City, as a megalopolis in which technological progress, social decadence and ecological disasters have made life practically impossible" (1). I agree with de Fays' ecocritical conclusions about Moctezuma City, but I think Aridjis paints a much more complex picture of responsibility for ecoapocalypse.

Throughout the novel, Aridjis links human alienation to a paucity of environmental health and artistic expression. For example, the drama troupe with which Yo is associated often performs works from Golden Age Spain. Faced with the death of the theater, one of the novel's thespian characters buries in the patio *La Celestina* (Fernando de Rojas), *Antigone* (Sophocles), *The Tempest* (William Shakespeare), *Bacchantes* (or *Bacchae*, Euripides) in hopes that they will grow to life in the future (56). While other city residents blink mindlessly at automated chirping birds in metallic trees, the aging actress Arira, along with her sisters, is one of few with the intuition that new life springs from the earth.

With such scenes, Aridjis signals a connection between art and ecological health and a negative association between consumer culture and environmental degradation. According to Stephen Mosley, environmental historians are now beginning to examine how "patterns of mass consumption (both material and cultural), as well as modes of production, can play a significant part in driving ecological and societal changes" (927). In a 1998 interview with Rubén Don, Aridjis, too, noted the loss of place-based referents in a consumer-driven megalopolis:

> encontrarme a las seis de la tarde en periférico o en el viaducto es estar flotando en la nada, en una nada, es como un baño de nada, es puro coche, puro tráfico, contaminación, ruido, aire.... Entonces todas esas

multitudes espontáneas que se hacen en todas partes de la ciudad, todo eso es también la experiencia ecológica que me ha servido (5). [to find myself at six in the evening on the beltway or viaduct road is to be floating in nothingness, in a nothingness, it is like a bath of nothingness, only cars, only traffic, pollution, noise, air. . . . Then all these spontaneous crowds that form everywhere in the city, all this has also been an ecological experience that has informed me.]

Modes of production and patterns of consumption produce traffic jams, dense smog, water shortages, and miserable living conditions of Mexico City, and Aridjis himself counts these as part of his environmental education.

Aridjis makes his protagonist Yo the figure in whom biophilia persists in the midst of ecoapocalypse. Consider, for example, this passage narrated by Yo as she tells of her encounter with a giraffe in the Ecological Conservation Park: "El caso es que ante la presencia de mi doble natural sentí vértigo y apreté los párpados. Oí su voz, semejante a un ronquido, pero no comprendí lo que me dijo" (87) [What happened is that standing before my natural double, I felt dizzy and squeezed my eyes shut. I heard her voice, like a snore, but I did not understand what she told me]. A crowd gathers, and people begin to stare at Yo and the giraffe, noticing the resemblance. Yo leaves annoyed at the crowd and saddened by the solitary animal: "me alejé de la multitud, volviendo la cabeza sobre el hombro para ver por última vez a ese animal sin pareja, el último de su especie, que se extinguía lejos de su hábitat" (88) [I moved away from the crowd, looking back once over my shoulder to see this animal without a mate, the last of its species, dying out far from its habitat]. This passage serves to emphasize that Yo, like the giraffe, lives far from her ideal habitat, solitary and awkward among gawking bystanders.

She is a misfit in Moctezuma City, but in the Nahua cosmology that undergirds the novel, Yo is well suited to be the harbinger of a "new race" of humans, better than the last. According to Nahua traditions, there have been several creations and cataclysms in the history of the world. According to renowned Mexicanist Miguel León-Portilla:

In Nahua thought, the world had existed not once, but several times. What was called 'the first foundation of the world' had taken place many thousands of years ago; so many that four distinct ages, called Suns by the ancient Mexicans, with their four different universes, had existed prior to the present epoch. In those Suns there had been a certain "spiral" evolution in which progressively better, more complex forms of inhabitants, plants and food had appeared (4).

Each of the ages ended with a cataclysmic event, but there were always survivors. According to León-Portilla, "since each age, or Sun, always ended with a less than complete cataclysm, instead of history returning to repeat itself in a fatally identical manner, every new cycle ascended spirally, giving rise to better forms of life" (4–5). Out of the ashes or ruins of each previous epoch came an age that was better: four total, and humanity presently inhabits the Fifth Age.

Nahua cosmology allows Aridjis to offer humans in his story a way out of extinction.[4] When Aridjis anchors his futuristic novel to a non-Western conception of time, he places ecocide in the megalopolis in a larger, more-than-human, geologic timeframe. This displaces the focus of the story; no longer is it the tale of modern, human society alone. Contemporary human dramas are part of a much larger context of natural history and the more-than-human world, as the Aztecs and their predecessors understood.

In the last pages of the novel, auguries of the end of the age increase. As tremors shake the city, Yo appears to find her mate, a giant man named Baltazar whose presence is associated with the tremors and quakes. Their first sexual encounter is unpromising; Yo is bored and distracted, and the moment is emblematic of the alienation in human society. Nevertheless, Yo later finds herself thinking of Baltazar. They part when he informs her that he needs to leave the city for his health. His absence leads Yo to desire his presence again, to think of him and long for his return. Until then, Yo has been one more distracted lover in Moctezuma City, a fact to which the title of the novel alludes. Aridjis presents a physical and psychological connection between two misfits, Yo and Baltazar, as the conduit for new possibilities amid the alienation in Moctezuma City.

In the last moments of the novel, Baltazar appears again, this time with a blue and gold standard proclaiming the end of the Fifth Sun. He tells Yo he has come for her and asks her, "¿En quién piensas cuando haces el amor?" (272) [Who do you think about when you make love?] She answers that she thinks of him. If Yo and Baltazar connect on an intimate level, the possibility of life is restored for human beings. Sex is the center of the reproductive cycle and the cycle of life. Thoughts of union accompany the physical union of Yo and Baltazar, so real psychological intimacy is also restored. Aridjis presents the mind and body of his protagonist united in one creative, procreative purpose. The reproductive cycle of life, center of biology and theme of countless myths, is key to the rapid denouement of the plot.

Aridjis ends the modern nightmare of Moctezuma City with a devastating earthquake that demolishes it. The human-constructed pseudo-environment rests on a geologic reality that residents cannot eliminate or escape. Like the floods in other novels, the earthquake is a cataclysmic event. With this ending, Aridjis reminds readers that human societies are fundamentally connected to the earth, whether they wish to acknowledge this or not. Misfit Yo becomes a hopeful figure, the one in whom a new future may be established upon the ruins of the Quinto Sol. What are liabilities in 2027—freakish height, irrepressible biophilia, and historical consciousness—make her the perfect figure to start a new age. As Baltazar and Yo are reunited in the ruins of Moctezuma City, the birds begin to sing "creyendo que era el alba" (273) [thinking it was the dawn]. According to Aztec cosmology, the birds are correct.

Waslala

Just as Aridjis reaches back into Nahua cosmology, Gioconda Belli also draws from the heritage of her country in search of a route out of apocalypse. To depict hope for the futuristic world in *Waslala*, Belli draws inspiration from the rich poetic tradition and heady revolutionary dreams of her native Nicaragua.[5] Belli's Faguas exists as a failed state, linked to the global economy in a neocolonial relationship with wealthy nations.

Like Aridjis, Belli is well known in the field of Latin American letters for both her writing and public profile. She is most famous for her militancy during the Sandinista era and for her outspoken feminism; she brilliantly combined the two in her most successful narratives: the immensely popular novel *La mujer habitada* (1992) and her memoir *El país bajo mi piel* (2001). Despite her success in narrative, Belli began her literary career writing poetry, crafting an erotic, woman-centered art with a social message beginning in the 1970s. During the revolution, she rebelled against her bourgeois upbringing by joining the Sandinistas in the fight against the Somoza dictatorship and also by ditching her traditional marriage for a series of lovers and partners. Cultural emissary for the Sandinista government in the 1980s and winner of numerous literary prizes, Belli is among the most widely read contemporary Central American authors.

Most literary criticism has focused on Belli's feminism rather than her depiction of place. *Waslala* is similar to Belli's earlier works in its preoccupation with Faguas, but the novel stands out from an ecocritical perspective because it brings environmental destruction into the foreground of a long history of social injustice. In *Waslala,* Belli explicitly links environmental degradation to the exploitation of the people of Faguas. She also posits two sites of resistance and regeneration: an erudite, poetic tradition, intimately tied to place, and a colloquial, creative ingenuity, rooted in poverty.

The plot concerns the journey of protagonist Melisandra into the interior of Faguas in search of the lost utopian community of Waslala. Young, comfortable, independent-minded Melisandra sets out from her grandfather's riverside hacienda on a physical and metaphorical journey into self-knowledge. The catalyst for the trip is the arrival of a band of foreign visitors, among whom is a journalist, Raphael, who asks her to guide him into the interior. On the journey upriver, Melisandra joins an eclectic group of foreigners, including a lesbian couple, an arms trafficker, and a scientist. Melisandra and the reporter, in particular, hope to find Waslala; the others visit Faguas for their own reasons.

Waslala offers challenges to ecocritical predilections on a number of levels. For example, though Belli is a feminist writer with an ecological consciousness, I do not see *Waslala* as an ecofeminist text because it perpetuates certain gender stereotypes and lacks a clarity of metaphor about the nonhuman, natural world. Yet it would also be problematic to dismiss the novel outright on ecocritical criteria. Muddy metaphors and all, Belli articulates an approach to environmental issues that speaks from an important place, marked by difference. And, as Greta Gaard notes, "perhaps the first step in developing a cross-cultural ecofeminist ethics is to acknowledge the largely Western cultural and economic contexts in which ecofeminism has been developed" ("Strategies" 83).

Reconsidering ecofeminism from a more inclusive perspective presents multiple challenges, and Belli's book illustrates these well. Gaard suggests that "an ecofeminist approach to developing a cross-cultural ethics would require listening to the experiences and analyses of the women working for social and environmental justice within a particular culture, building relationships among cultural insiders and outsiders, and working in solidarity to support the insiders' efforts" ("Strategies" 84). Gaard posits an impressive list of considerations, to which I would add the need to recognize the multiple allegiances and identities even of those cultural "insiders" like Gioconda Belli. As Nicasio Urbina has pointed out, part of Belli's success in popular works like her memoir is attributable to her mobilization in narrative of all the social capital she has at her disposal (Urbina). Her memoir, in which Belli catalogues the allegiances and internal debates of her own life, itself points toward the danger of assuming that identity politics is absent from territories of the famous "Other" of literary fame. To see any one author as a spokesperson for a Latin American (or even Central American or Nicaraguan) environmentalism or ecofeminism is naïve. Belli's perspective differs widely, for example, from that of Anacristina Rossi and Tatiana Lobo, even though all three articulate feminist positions and use a rhetoric of nature to problematize contemporary realities in Central America.

Waslala is typical of Belli's narrative in that it communicates a will to power through the protagonist. It does not posit nonhierarchical power relations, nor does it bring into the foreground of debate voices from different ethnic groups in Latin America. Belli does not advocate radical new power relationships; she critiques the current world order and posits a relatively traditional solution through her feminist protagonist. Indeed, Belli does not even offer a protagonist who takes her place among the inhabitants, human and nonhuman, as Rossi's Daniela did in *La loca de Gandoca*. In *Waslala*, the benevolently powerful confront those who are malevolently so. Melisandra emerges as the leader, bolstered by the popular assent of affectionate, well-meaning masses, and she shows tendencies toward benevolent, authoritarian rule.

In terms of ecological imagination at the level of metaphor, Belli's novel is mixed. At times, Belli's language shows that she has inherited a rich legacy of Nicaraguan poetry with deep, local knowledge of the natural world. At other times, Belli uses tropes that are traditional, trite, and even consumerist. In some instances, the metaphorical language explicitly commodifies nature. Consider here opening lines in the novel, indicative of both the possibilities and limits of Belli's ecological imagination:

Era una lástima saber que cuando se fuera no podría llevarse el río anudado a la garganta como una estola de agua. Le era difícil imaginar la vida sin aquel caudal cuya tumultuosidad o mansedumbre, marcaba las estaciones, el decurso del tiempo (13).

[It was a shame to know that when she left, she wouldn't be able to wear the river around her neck like a stole of water. It was difficult to imagine life without that flow whose tumultuousness or tranquility marked the seasons, the passage of time.]

The passage describes the river upon whose banks Melisandra lives with her grandfather-poet, a figure Belli modeled on Nicaraguan poet José Coronel Urtecho (1906–1994). Urtecho was a leader of the Vanguardists and a mentor to many younger poets, including Belli herself. The significance of Urtecho and other Vanguardists, including Pablo Antonio Cuadra, in the creation of a Nicaraguan environmental imagination is undeniable, and *Waslala* is in many ways a tribute to that poetic tradition. Nevertheless, the opening passage falls short of Urtecho's legacy. It signals the spiritual value of the river, but unfortunately it uses a metaphor that posits the natural world as fashion accessory. A woman replaces a man in the space of desirer and possessor, but the metaphor still conveys a very anthropocentric function for the natural world.

Belli's ecological imagination is distinct from the "deep" ecological consciousness that authors like Rossi and Sepúlveda articulate through their protagonists.[6] However, *Waslala* has other merits. It is particularly effective in making environmental destruction and social chaos visible and in linking the devastation of landscapes to neocolonial patterns promulgated by globalization.

In *Waslala*, Belli challenges readers to apprehend a global ecological apocalypse that is felt differently depending on the geopolitical position of the human being and environment in question. She contrasts modernity in the developed and developing worlds in stark terms, and she articulates an "environmentalism of the poor" (Guha, *Environmentalism* 102). According to Guha, there are several characteristics of the environmentalism of the poor. He notes that "first and foremost, it combines a concern for the environment with an often more visible concern for social justice" (*Environmentalism* 105). In Faguas, environmental justice issues abound: dumping of consumer goods in the developing world; toxic and radioactive contamination; forest destruction; mass cultivation of a hybrid plant for illegal narcotics; global environmental inequity. Many suffer, yet few are able to challenge this injustice, and those that do are predominantly women. In Belli's novel, the active, world-changing characters are women, principally Engracia and Melisandra (and also to a degree Melisandra's mother). Interestingly, the prominence of women in the cause of environmental justice is another of Guha's characteristics of the environmentalism of the poor (*Environmentalism* 109).

Belli's futuristic world, like that of Aridjis, is a failed state in which community life has all but disappeared. Even within families, life is in disarray. Women mourn their dead; national governments are obsolete; and regional, extremist struggles flare into violent confrontations. Belli dramatically represents the nightmarish ways in which ordinary, impoverished people live and die in a web of power relations that rip the human and nonhuman environment to shreds. While other regions enjoy prosperity and leisure, Faguas trades in oxygen and drugs, receives toxic waste, and traffics all manner of contraband for capital to fund its perpetual fratricidal wars.

Belli's vision of the developed world beyond Faguas is equally dismal. There, technology has led to a definitive divorce between humans and nature. Belli's representation of environmental degradation in modernity emphasizes the prosperity of the few abroad juxtaposed against the misery of the multitudes:

> Los contrabandistas se llevaban minerales y sabe Dios qué otras cosas y traían a cambio armas, lotes de mercancías caducas, artefactos, objetos que en Faguas eran codiciados porque, después de todo, a cierto adelanto se acostumbraron antes de que se les descartara y se les declarara insalubres, un virus maligno que amenazaba, con su mera existencia, la vida civilizada, avanzada, afanada ahora con la idea de la exploración espacial, de emigrar en masa y empezar de nuevo en otra galaxia donde no se filtrara nunca por ninguna ranura la noción de otros seres humanos excluidos, subsistiendo en condiciones primitivas, míseras, reproduciendo sin su pobreza, sus guerras, sus plagas sin control (19).
> [The smugglers carried off minerals and God knows what other things, and in exchange, they brought guns, expired lots of merchandise, artifacts, objects that in Faguas were desired because after all, they were

useful before they were discarded and declared unhealthy, a malignant virus that menaced, by its very existence, civilized life, advanced life, life now enthused by the idea of space exploration, of emigrating en masse and beginning anew in another galaxy where there would never filter in, through any crack, any notion whatsoever of other excluded human beings subsisting in primitive, miserable conditions, reproducing, without control, their poverty, their wars, their plagues.]

Passages such as this one reveal the dangers of an ever-widening gulf that proclaims the endgame of modernization.

More than anything else, Belli's ecological imagination highlights what Ramachandra Guha and J. Martínez-Alier have called "ecological distribution conflicts" (31). These are "social, spatial and temporal asymmetries or inequalities in the use by humans of environmental resources and services . . . and in the burdens of pollution" (31). Belli links this reality in modern Faguas to a history of human and ecological exploitation initiated during the colonial era. Consider, for instance, this passage focalized through Don José:

En los ojos de los modernos navegantes, cuántas veces no vio él la codicia con que surcarían el río los filibusteros, los comerciantes, el comodoro Cornelius Vanderbilt. . . . Río abajo, río arriba viajaban los extranjeros cargando delirios de grandeza, sueños, quimeras de canales interoceánicos, mitos de lo que se podría hacer con ese país si sus habitants se traicionaban los unos a los otros (18).

[How many times did he see in the eyes of modern sailors the same greed with which the others plowed through the river, filibusterers, traders, Commodore Cornelius Vanderbilt. . . . Downriver, upriver the foreigners traveled carrying delusions of grandeur, dreams, chimeras of interoceanic canals, myths about what one could do with this country if its inhabitants would betray one another.]

Belli resignifies Faguas' famous river (the Río San Juan in Nicaragua) as a living symbol of a legacy of exploitation, war, and resource extraction that continues uninterrupted into the dystopian future.[7]

Belli's novel sets up a polarized world in which Faguas is the site of human misery in the midst of natural wealth, while the powerful, developed

world is a prosperous but sterile wasteland. The sort of geopolitical disparities Belli critiques have recently drawn the attention of foundational figures
in ecocriticism. For example, Lawrence Buell cites *maquiladoras* (assembly
plants) along the U.S.-Mexican border and barges of waste sent to Africa
as industry responses to environmental activism in developed countries
("Toxic Discourse" 644). According to Buell, such moves are "aggravating
the problem of eco-inequality on a global scale and approaching, it would
seem, the dystopian end point of modernization" ("Toxic Discourse" 644).
Belli puts the discourse of nature in the service of a critique of this reality.
Her imagined dystopia lays bare the fragmented, hierarchal world that, for
her and for Aridjis, lies just beyond the latest phase of modernization.

Belli also critiques some strains of first-world environmentalism. In particular, she condemns the hypocrisy of an international community whose
advocacy of environmental issues is linked purely to self-interest. For instance, in Faguas the "Environmental Police" patrol protected forests, wor-
ried more about oxygen supply than the masses of human beings who suffer
from violence and desperate poverty. For the developed world, Faguas is one
of the

> manchas verdes sin rasgos, sin indicación de ciudades: regiones ais-
> ladas, cortadas del desarrollo, la civilización, la técnica, reducidas a
> selvas, reservas forestales, a función de pulmón y basurero del mundo
> desarrollado que las explotó para sumirlas después en el olvido, en la
> miseria, condenándolas al ostracismo, a la categoría de terras incog-
> nitas, malditas, tierras de guerra y epidemias adonde sólo llegaban los
> contrabandistas (19).
>
> [green stains with no defining characteristics, no hint of cities: isolated
> regions, cut off from development, civilization, technology, reduced
> to jungles, forest reserves, lungs and garbage dump to the developed
> world that exploited them only to subject them later to the realm of the
> forgotten, of the desperately poor, condemning them to ostracism, the
> category of terras incognitas, cursed places, lands of war and epidemics
> frequented only by smugglers.]

The passage refracts with the perspective of residents of the underde-
veloped world, who see themselves as they are seen by developed nations.
Though problematic for nature-first environmentalists, the passage docu-

ments an important tension in environmental struggles. Until recently, there has been a rift between the environmentalism of the world's impoverished peoples and the "environmentalism of the affluent" (Sheppard 135). When affluent environmentalism privileges nature-first concerns while disregarding the plight of impoverished people, it loses potential allies among vast populations in the developing world.

Belli imagines and then critiques a situation in which residents of the developed world are separated from nature in their daily lives and worry about wilderness only when its disappearance imperils their existence or pleasure. With such representations, Belli critiques modernization on various levels. First, Belli points out that impoverished countries cannot improve living conditions if neocolonial models remain unchanged. Neither can impoverished countries aspire to levels of affluence in wealthy nations because these are not ecologically sustainable at a global level. So, when wealthy nations impose conservation abroad without also cutting consumption at home, they create a dilemma for poor countries: choose conservation and with it, poverty; or opt for development, and incur the wrath of the developed world for tipping the earth toward ecological apocalypse.

In this way, Belli takes current economic rifts, and their attendant environmental effects, and projects them forward toward logical extremes. At the same time, Belli also cautions that saving the natural world is a project fraught with difficulties. In *Waslala* actions "for the greater good" taken by international governing bodies have unforeseen consequences, at least in part because planners have an inadequate grasp of the realities of desperately poor regions. For instance, in Faguas the protection of the forests by international forces bolsters an arms trade that has had the region embroiled in constant wars. Consider this wry comment, in the voice of an arms trafficker:

> Hay un nuevo brote de terrorismo ecológico. Los terroristas le prendieron fuego a varias hectáreas de bosque, pero la Policía Ambiental, con sus helicópteros, logró apagarlo rápidamente. Esto afectará, sin embargo, los próximos convenios. Cuando vengan los ejecutivos de la Corporación del Medioambiente exigirán patrullas armadas de guardabosques. . . . Eso es bueno para mí, por supuesto (110).
>
> [There is a new outbreak of ecological terrorism. The terrorists set fire

to various acres of forest, but the Environmental Police with their hel-
icopters managed to put it out quickly. However, that will impact the
next agreements. When the executives of the Environmental Corporation come, they'll demand armed forest patrols. . . . That's good for me,
of course.]

Here, ahead of a new round of environmental accords, terrorists have
attacked forest preserves. The next agreement, then, will surely include pro-
visions for armed forest rangers, a boon for the arms trafficker who benefits from both legal and illegal arms trading. Though few environmentalists would question the value of protecting forest reserves, Belli shows that
the apparently simple action of forest conservation can have multiple social
ramifications. In this case, one consequence is the escalation of violence felt
by the particular sector of humanity that has the least voice in decision-
making and yet most feels the impact of environmental agreements.

Not all in *Waslala*, though, is a condemnation of great global inequalities.
Belli celebrates leadership by strong women, art that promotes hope, and
ingenuity that permits survival. There are three strong women in *Waslala*:
Melisandra's mother, Engracia, and Melisandra, each of whom offers a site
of resistance and a hope for regeneration. In contrast, men, particularly the
Espada brothers, control the business of war and the trafficking of contraband; they wield the power of destruction and death. Men who are dissent-
ers gravitate toward strong women: Don José to Melisandra's grandmother,
Morris to Engracia, Raphael to Melisandra.

For her part, Engracia builds life in collaboration with the ingenuity of
Faguan workers. Belli has Engracia's power emanate from a mountain of
refuse. Under her magnanimous leadership, the dump community in Cineria stands as a strange refuge, a recycling center of the impoverished world,
where people sort through containers of trash to give useful items new life.
An emigré from Waslala, Engracia is a woman larger-than-life; her imposing stature, good will, and honesty inspire the solidarity of the community
around her, despite the unlikely setting.

In Engracia's practical dump community live the dreams and memories of
Waslala, of the impractical utopia. Cineria itself is a city on the banks of a lake,
where "restos de cuanto objeto cupiera en la imaginación yacían . . . apillados
en grandes montañas" (133) [remains of every imaginable object lie . . . piled

in huge mountains]. An outcast and misfit herself, Engracia has built an economy out of the reuse and rehabilitation of discarded consumer products that arrive in the barges of refuse. In their midst, Engracia has staked out a small territory that has an uneasy, defensive peace with the Espadas. Cineria dramatizes the impact of consumer society on those who receive its waste, but Belli also dignifies the dump. She populates it with individuals, highlights its structure, and casts it as a center of alternative power in Faguas.

Belli also uses Cineria and Engracia to deliver a message of environmental justice. Ultimately, the shipping containers hold the seeds of death for Engracia and her scientist-lover Morris. Delighted with the phosphorescent blue powder discovered in a container, Engracia and several workers adorn themselves with the substance only to find that it is Cesium 137, a radioactive isotope (184–85).[8] In an act of self-sacrifice, Morris casts his lot with Engracia and adorns himself, too. Aware of their imminent death, Engracia and Morris mount a suicide attack on the Espada offices, figuratively and literally demolishing the old order and thereby making way for Melisandra to create a new society.

At the conclusion of the novel, Melisandra finds and then leaves Waslala. She discovers the place thanks to a local guide and to Engracia's parrot, both of which guide Melisandra's trek through a forest familiar to them. Once in the community, Melisandra discovers it has been abandoned by all but her mother. Her mother explains that the population dwindled and the poet-founders died, but that those that remained realized that Waslala served a symbolic, inspirational function for the Faguan population. And so they dedicated themselves to making a spectacular place of beauty and hope: cultivating gardens, guiding the branches of the forest canopy, channeling the murmuring brook. When Melisandra relates that though no more seekers come, no one in Faguas has forgotten Waslala, her mother responds thus:

Es la memoria, Melisandra. Siempre pensamos que la memoria debe de referirse al pasado, pero es mi convicción que hay también una memoria, un memorial del futuro; que también albergamos el recuerdo de lo que puede llegar a ser (329).

[It's memory, Melisandra. We always think that memory should refer to the past, but it's my belief that there is also a memory, a memorial, to the future; that we also harbor the memory of what could come to pass.]

With these lines about Waslala, Belli memorializes the utopian dreams of the Nicaraguan revolutionary movement in which she herself came of age as a woman, citizen, and writer. She also redeems the notion of a utopian impulse in revolutionary movements and signals a role for art in shaping a better future.

In Waslala, Melisandra finds the missing link in her own history (that of her mother), and she rejects the sterility of a society built in isolation from the problems of the world. With this image of Waslala, Belli vindicates the place of utopian dreams, and she signals the important role of poets and practical women in inspiring movements for justice. Poets record important links between humans and more-than-human nature, and doers like Engracia can carve a better reality out of dismal surroundings. Don José articulates a passion for the land in which he lives in his appreciation for poetry, but the two lovers in his life (Engracia and his wife) build oases of resistance and sustainability, one in a dump and one on the banks of a river. This practical, matriarchal legacy will converge in Melisandra, who must work to create the reality Engracia imagines: "un mundo bienaventurado en donde . . . ni yo ni tantos y tantos tengan que morir y vivir entre los desechos y despojos" (286) [a blessed world where . . . neither I nor so many, many others would have to live and die among waste and remains].

As place and as hope, Waslala lies at the center of Melisandra's quest for self and a sense of place. Elusive and unreachable in poetry and popular lore, the myth of Waslala brings hope to thousands of poor Faguans. For them Waslala is a reality, a memory whose counterweight compensates for the daily violence of their poverty and the ruthless control of a drug-trafficking oligarchy.

Belli casts Waslala in its prime as an agrarian, pastoral paradise, steeped in poetry and sited in a valley teeming with life and beauty (244). In its environs, the human hand in nature has altered the local ecosystem in much the way anthropologists like Darrell Posey have described: increasing its biodiversity and its beauty for humans (312). According to Don José, the poet Ernesto was a driving force behind the vision for Waslala:

Ernesto nos habló de un sitio en el Norte del país. Nos describió su gran belleza natural, el arroyo que lo atravesaba, las montañas circundantes que crecían alrededor de la selva tropical magnífica. Era sitio donde los sueños adquirían texturas vívidas y fantásticas. Durante una de las noches que él pernoctó allí soñó con una ciudad plateada. Su nombre, 'Waslala', aparecía resplandeciente sobre los troncos viejos y monumentales de los ceibos (53–54).

[Ernesto spoke to us of a site in the north of the country. He described to us a place of great, natural beauty, the river that ran through it, the encircling mountains that rose up around the magnificent tropical forest. It was a place where dreams took on vivid, fantastical textures. During one of the nights he spent there, he dreamed of a silver city. Its name, Waslala, shone above the monumental old trunks of the ceibas.]

In the above passage, Belli references the "El Dorado" myth (changing the vision to one of a "ciudad plateada"), but she also references a poetic tradition that reaches back into Nicaraguan Vanguardism. Vanguardism flourished in Nicaragua in the late 1920s and into the 1930s, with Granada as its artistic center. Influenced by French and Spanish poets, as well as North American writers like Carl Sandburg and Walt Whitman, Vanguardist artists rejected bourgeois commercialism, extolled agrarian and campesino traditions, and advocated a nationalist literary expression (Arellano xvi-xvii).

Belli's work echoes Vanguardist nationalism, and their rejection of commercialism, but with the reference to ceibas, Belli deeply roots Waslala in ancient Mesoamerican cosmologies. Her work thereby grounds Waslala in deeply seated beliefs about trees, time, and the order of the universe, just as Aridjis anchored Moctezuma City to pre-Columbian epistemologies. The ceiba tree (*Ceiba petandra)* is, according to the *Neotropical Companion*, "one of the commonest, most widespread, and most majestic Neotropical trees" (Kricher 73). Reaching fifty meters in height, with buttressed roots, it is important to Mesoamerican indigenous cultures and is the sacred tree of the Mayans. When Don José leaves Waslala and tries to return without success, he attributes his failure to the disappearance of the ceibas and to the world they anchor:

En su mitología, que proviene de raíces mayas y aztecas, la ceiba es un árbol sagrado, el árbol que sostiene el mundo; si desaparace la ceiba, el mundo que sostiene desaparece con ella (57).

[in their mythology, drawn from Maya and Aztec roots, the ceiba is a sacred tree, the tree that supports the world; if the ceiba disappears, the world it supports disappears with it.]

Passages like this signal a substrate of indigenous environmental consciousness that enriches Central American oral traditions. These beliefs about the place of human societies amid the more-than-human world permeate the poetry of some of the best writers from the isthmus.[9]

Belli suggests that despite this rich poetic legacy, or maybe because of it, Waslala fails. In its extreme, the utopian poetic tradition in Waslala produces the same effect as the picturesque tendency of "landscape arranging" Jonathan Bate discusses (119–52). Obsessed with beauty and peace, Waslala's authors create artifice instead, and it stifles life. When Waslala collapses, Melisandra's mother sacrifices her own life to keep its hope alive for others. The utopian, poetic tradition to which she is caretaker nourishes Faguas, but the legacy alone is not sufficient to pull the country out of war and into life.

The novel suggests that a new sustainable society will be built upon art and ingenuity, two legacies that Melisandra inherits. Having confronted her own past and experienced the reality of the interior, Melisandra emerges a strong, centered woman. From the maternal figures she comes to know in her quest, Engracia and her biological mother, Melisandra finds a model for practical action against the geopolitics of inequity. From the men in her life, namely her grandfather and Raphael, with whom she indulges in a budding romance, she finds inspiration in the truths of poetry.

More than any of the other novels considered here, *Waslala* imagines an engagement with the complexities of the world, of Faguas, and of globalization. Melisandra consciously rejects a retreat from modernity, and she embraces an ambiguous future. Melisandra's exit from Waslala is a gesture of practicality, and with it, Belli suggests the need to confront the challenges of modernity. Stirred by poets who express both biophilia and a deep respect for people, her protagonist sets out to affirm the right to dignity and life for Faguas. With her imagination anchored in Waslala and the river, Melisandra welcomes the chance to prevail in Faguas, where a culture of death and degradation have long held sway.

Conclusions

Belli's novel points ecocritics toward the need to consider both poetic legacies and global inequity when they consider literary texts. Her epigraphs suggest that the task of environmental justice is unending, ethically compelling, epic, and difficult. One epigraph cites a passage from Alfred, Lord Tennyson's *Ulysses* that begins "Come, my friends. / 'Tis not too late to seek a newer world" and concludes with the line "To strive, to seek, to find and not to yield" (9). The next epigraph is a quote Belli attributes to an unnamed Tanzanian president: "Hay quienes quieren llegar a la luna, mientras nosotros aún estamos tratando de llegar a la aldea" (9) [There are those who want to reach the moon, while we are still trying to get to the village]. With *Waslala*, Belli insists that great, global divides are fundamentally significant in attitudes toward nonhuman nature, and they must be recognized by environmentalists.

As ecologically conscious responses to globalization, both Aridjis' and Belli's novels are instructive narratives. They offer visions of Latin America grounded in a unique history and place. And like all the authors in this study, Belli and Aridjis ask readers to comprehend the place of the collective humanity in Latin America, amid nonhuman nature and in the geopolitical world order at the end of the century.

Coda

New novels of ecological imagination have appeared in a moment of rapidly accelerating globalization in Latin America. During this period, legacies of neocolonialism, corruption, and authoritarianism still confronted Latin American peoples, even as they struggled to come to terms with new economic realities. Democracies took root, but at the same time, globalization weakened and challenged the prerogatives and responsiblities of the nation-state.

The novels we have explored here spring from a context of growing unease and disillusionment about the imbrication of Latin America in the world economy. The novels are important expressions of a dynamic ecological imagination alive in Latin American literature since its earliest days. Often in its literary history, authors have portrayed the natural world of Latin America as exuberant and inexhaustible. Writers of the Boom and magical realism emphasized the representation of Latin American space as both exotic and fertile. But in the last decades of the twentieth century, as cities boomed and forests fell, some authors shifted discursive preferences and adopted a more skeptical, confrontational, or elegaic tone.

New Latin American ecological imaginations instill in the minds of readers a world with real limits. They emphasize the boundaries imposed by the nonhuman world on human societies. They draw attention to both environmental degradation and unresolved social issues: land claims, indigenous rights, classism, sexism, and ethnic exclusion. And ultimately, I would argue, authors of ecological imagination from Latin America wield the discourse of nature as a literary weapon against the homogenizing agenda of neoco-lonial and neoliberal enterprises. They craft this discourse of nature in their fictions as an intervention for democracy, for local knowledge, for human rights and environmental justice. Theirs is an ecological imagination imbued with social justice, and their new story of Latin America represents a real engagement with an imperiled world.

Notes

Introduction

1. Neoliberalism is an imprecise term that refers to resurgence of classical liberal economic ideas. Neoliberal economic policies generally push for free markets and free trade and recommend breaking down barriers to trade and investment in order to achieve national economic prosperity. The term "neoliberal" is the same in Spanish as in English, and it is widely used in Latin America by critics of free-market policies.

2. Some titles are available in English translation. Of the list, English translations currently exist for the titles by Iparraguirre, Souza, Sguiglia, and Sepúlveda. There is an unauthorized English translation for the Rossi title, and an authorized version is reportedly forthcoming. When English translations are available, I have used the published translation, and indicate the page references parenthetically after both the Spanish quote and published English translation. All other translations are my own.

3. Bate uses the phrase to explain the persistent popularity in the Anglo world of works by Jane Austen and Thomas Hardy.

4. For more on the notion of "global designs" and their imposition on local epistemologies, see Walter Mignolo's *Local Histories/Global Designs: Coloniality, Subaltern Knowledges, and Border Thinking*.

5. See Glen Love's article "Revaluing Nature: Toward an Ecological Criticism" (1990) and Steven Rosendale's "Introduction: Extending Ecocriticism" (2002) for more on the shift in emphasis.

6. The same *Nature Conservancy* issue also featured articles by the likes of Jeffrey Sachs, a prominent economist and the head of the Earth Institute at Columbia University. In his article, Sachs argued that "there is an ever-present risk of a downward spiral in which environmental degradation worsens poverty and in which deepening poverty accelerates

environmental degradation" (28). He notes that the poor overexploit natural resources themselves because they cannot afford to do otherwise, and also that the poor are "vulnerable to manipulation from rich countries and powerful corporations, which often irresponsibly mine minerals, cut down forests and overexploit biodiversity to service rich-country markets" (28).

7. Candido's essay was originally published in French in 1970 (Del Sarto 28). I quote here from the English version published in *The Latin American Cultural Studies Reader*.

8. García Canclini uses the term "modernity" in the sense that it is understood from the writings of Theodor Adorno or Walter Benjamin. "Modernity" should not be confused with the Latin American artistic movement known as "modernismo," associated with poet Rubén Darío and other artists in the late nineteenth and early twentieth century. Unfortunately, the Spanish words "modernismo" (modernism), "modernidad" (modernity), and "modernización" (modernization) have all been translated and employed in different ways over time, and this leads to confusion about the connotations of the term. I use modernity in the sense of García Canclini, Adorno, and Benjamin and use the Spanish word "modernismo" for the artistic movement.

9. For an excellent, in-depth introduction to ecocriticism and Caribbean literature, see *Caribbean Literature and the Environment: Between Nature and Culture*. Ed. Elizabeth M. DeLoughrey, George B. Handley, and Renee K. Gosson.

Chapter 1. Alterity, Empire, and Nation in Tierra del Fuego

1. Demitrópulos' historical novel *Río de las congojas* (1981) was published in English as *River of Sorrows* (2000); *Piano*, though, is not available in English translation.

2. The Dirty War refers to the period of dictatorship in Argentina between 1976 and 1983. A campaign against dissidents and alleged terrorists by the ruling military junta resulted in the disappearance, torture, and death of thousands of civilians.

3. The Brazilian story of independence from Portugal is more complicated. Brazil actually became the seat of empire in 1822 when the Portuguese monarch Pedro I went into exile in the former colony. Brazil became a republic in 1889.

4. For further reading on the birth of the modern world system, particularly with regard to the role of the Americas in its creation, see, Aníbal Quijano and Immanuel Wallerstein's 1992 article, "Americanity as a Concept." There the authors argue that the creation of the "geosocial entity" known as the Americas was "the constitutive act of the modern world-system" (549).

5. This is the post-independence era for most Latin American republics and the Victorian era for Britain.

6. Historians such as Robert B. Marks argue that the emergence of the modern world system came much earlier with the rise of manufacturing in China. In his analysis, the Americas and Australia/New Zealand become inserted into a preexisting world system that thereby becomes truly global. For a succinct summary of his ideas, see his book *The Origins of the Modern World* (2002).

7. The Spanish made early attempts at establishing a permanent presence in the area.

Under the leadership of Sarmiento de Gamboa, a colonial outpost was attempted in 1584, but the settlement survived only a brief time. Colonists died in the harsh conditions, supplies and reinforcements never arrived, and only a handful of people survived to tell the story.

8. Argentina is a nation profoundly influenced by immigration. Government programs in the nineteenth century promoted the immigration of Europeans in massive numbers, and Argentine identity reflects the contributions of these immigrants. For more on the topic, see David Rock's *Argentina, 1516–1987*, particularly pages 118–61 on the formation of the nation-state.

9. For Darwin's account of the Button incident, see Chapter 10 of *Voyage of the Beagle* (1909).

10. Recent years have seen the publication, in English, of Harry Thompson's *This Thing of Darkness* (2005) and Nick Hazlewood's *Savage: The Life and Times of Jemmy Button* (2001), as well as a new biography of FitzRoy called *Evolution's Captain* by Peter Nichols. All these recount the Fuegian episode from different perspectives (none Latin American).

11. See Mignolo's *Local Histories/Global Designs: Coloniality, Subaltern Knowledges, and Border Thinking.*

12. Iparraguirre has said that Guevara was the key for her in writing the novel. Confronted with the history of Jemmy Button, she at first thought she would be unable to write a fictional work about it. Then, she said, "Cuando surge en mí el personaje de John William Guevara, que es inventado, recién sólo sentí que tenía la novela. . . . Guevara hace que la novela funcione" (Roffé 103) [When the fictional character of Guevara occurred to me, it was then that I knew I had the novel. . . . Guevara makes the novel work].

13. For more on the imperial gaze, see Mary Louise Pratt's book *Imperial Eyes: Travel Writing and Transculturation.*

14. Norman Cheadle comments at length on this resistance in Guevara's narrative (and in Iparraguirre's reconstruction of the Button affair) in his article about the novel.

15. In an interview published in 1997, Demitrópulos said that she based the character of Nancy on an actual woman. While reading a book by career military man José Otto Maveroff, a nineteenth-century envoy to Antarctica, she happened across the idea for the character. According to Demitrópulos, in his account of the journey, Maveroff described the predicament of an Englishwoman he met in a bar in Punta Arenas. She was playing the piano there to save money to return home, after having been deceived into immigrating to Argentina to marry a man who paid to have a bride sent to him (Castro and Jurovietzky 67).

16. The Mothers of the Plaza de Mayo is an organization that was formed by mothers of people disappeared during Argentina's military dictatorship (1976–1983), a time also known as the Dirty War. Each Thursday, these women marched (and continue to march), some holding photos of their loved ones, petitioning the government for answers about the fate of their relatives.

17. Belgrano Rawson states that in the novel "no hay ningún dato referencial; ni siquiera figuran el nombre 'Tierra del Fuego,' mucho menos las tribus fueginas, ni el Estrecho de Magallanes, ni el Cabo de Hornos; sin embargo, nadie puede tener dudas sobre cuál es el

lugar en el que se desarrolla" ("Escribir de Oído") [There is no referential information; not even the name "Tierra del Fuego" appears, much less the Fuegian tribes, Magellan Strait, or Cape Horn. Nonetheless, no one could possibly have any doubts about the place where the novel takes place].

18. For an excellent analysis of memory in the novel, see "Memorias de un mundo perdido: rememoración y evocación en *Fuegia* de Eduardo Belgrano Rawson" by Magdalena Perkowska-Alvarez.

Chapter 2. Contests for the Amazon

1. All three authors have drawn a substantial international audience, and the novels I discuss in this chapter are all available in English translation.

2. In *Practical Ecocriticism* (2003), Glen A. Love comments on the "aggressive anti-anthropocentrism" of his early ecocriticism and acknowledges other ecocritics like Steven Rosedale who argue there is room for a human focus in environmentally oriented criticism. Love agrees that "we have to keep finding out what it means to be human" and goes on to assert that the "key to new awareness resides in the life sciences" (6).

3. In an interview published in 2003 in a cultural supplement to the *Clarín* newspaper in Argentina, interviewer Sebastián Campanario actually carried out the interview in Sguiglia's offices at the Foreign Affairs Ministry.

4. For more on Sepúlveda's biography, as well as his take on trends in contemporary Latin American literature, see Andrew Graham-Yool's article "Light at the End of the Tunnel."

5. Slater argues that outsiders have interpreted the Amazon as various kinds of "giants": El Dorado, a Green Hell, the Rain Forest (13–14). Meanwhile, Amazonian dwellers—indigenous, black, and those of other cultures and mixed ethnicities—tend to represent Amazonian nature as a "shape-shifter." Their stories feature gold as a woman who changes appearance, the city of a lake within a lake, enchanted dolphins, and so on.

6. In *Ecology of Power* (2005), Michael J. Heckenberger presents a historical ethnographical study of the Xingú area in which he presents evidence of highly organized cultures in Amazonian regions other than the floodplains.

7. The railroad project was referred to as the Madeira-Mamoré project because it allowed for overland transport of freight and passengers between the Mamoré River and the Madeira River, thereby bypassing the falls on the Madeira River. The Mamoré River flows northward, delimiting the border between Brazil and Bolivia. It then joins several other rivers to form the Madeira, the longest tributary of the Amazon; the upper portion of the Madeira has a series of rapids and falls that make river transport hazardous. After traveling approximately 2,100 miles through modern Brazilian states of Rondônia and Amazonas, the Madeira joins the Amazon below Manaus.

8. Marcos' essay was originally published as "La 4e guerre mondiale a commencé" in *Le Monde diplomatique* in August 1997. The cultural studies journal *Nepantla* later published an English translation of the essay.

9. For more on the inspiration for the novel, see the interview published by Mario

Camara. In this interview, Sguiglia notes that a Brazilian historian contacted him after the novel's publication to congratulate him on representing a world he had known as a boy. Another traveler to Fordlandia told him that a boat-taxi operator told him that he would "get rich" thanks to "an Argentine that wrote about our history."

10. The term *caboclo* in Brazil generally refers to riverine people of mixed European and indigenous descent in the Amazon basin (Despres 14).

11. Among the best commentaries is that published by Camilo Gomides and Joseph Vogel in *Ometeca*. Another solid reading of the novel, in the context of Latin American writing about "la selva" is that of Rodrigo Malaver Rodríguez. Other researchers have used the novel as a springboard for discussions of contemporary environmental issues and feature a more simplistic analysis, as is the case with Daniuska González's "Viaje a la narrativa de Luis Sepúlveda."

12. For more on current conservation and environmental education efforts in the province of Zamora (through which the Nangaritza flows), see Kathryn R. Lynch's doctoral dissertation "Environmental Education and Conservation in Southern Ecuador: Constructing an Engaged Political Ecology Approach."

13. In 1960, Ecuador reneged on the Rio Protocol, signed in 1942 to end the border dispute between Ecuador and Peru.

14. For an excellent history of the transformation of the Ecuadorian Amazon, particularly with regard to the indigenous populations there, see Allen Gerlach's *Indians, Oil, and Politics* (2003). Chapter 4 is particularly relevant.

15. The Spanish word *tigre* can refer to any number of wild cats. The *tigre* of Sepúlveda's story is likely either a jaguar (*Panthera onca*) or ocelot (*Leopardus pardalis*), two of the larger cats native to Ecuador. Critics Camilo Gomides and Joseph Vogel translate *tigre* in their analysis as an ocelot.

16. Sepúlveda's assertions about a long history of Shuar aquaculture are backed by new, scientific evidence. Hugh Raffles and Antoinette M.G.A. WinklerPrins published in the *Latin American Research Review* a study of the manipulation of fluvial systems by people living in the Amazon. This study joins an increasing number of research projects pointing out the human interaction with and management of Amazonian "nature." This research is important, because as Raffles and WinklerPrins point out, much scientific and nonscientific literature throughout history has denied Amazonian peoples' agency by locating them in a "society of nature" (167). In many cases, geographic determinism informed this literature, producing caricatures of Amazonian character that had been formed by the formidable landscape. Instead, as scientists and other scholars point out, Amazonians have for ages been manipulating landscapes. Indeed, Raffles and WinklerPrins cite the studies of William Balée (14) that estimate that 12 percent of Amazonian forest is of biocultural origin (167).

Chapter 3. Paradise for Sale, or Fictions of Costa Rica

1. I use the term transculturation in the sense that Mary Louise Pratt employs it to refer to the ways in which "subordinated or marginal groups select and invent from

materials transmitted to them by a dominant or metropolitan culture" and "determine to varying extents what they absorb into their own, and they use it for" (6).

2. Foreign countries have shaped Central American political and economic development with profound social and environmental consequences. After independence from Spain in the 1820s, Central America fell under the imperial shadow of Great Britain. By the dawn of the twentieth century, the United States had eclipsed the British empire in hemispheric affairs and would dominate both politics and foreign investment into the twenty-first century.

3. For a thorough discussion of how one export commodity, coffee, altered Central American societies, states, and landscapes, see Robert G. Williams' *States and Social Evolution: Coffee and the Rise of National Governments in Central America*.

4. In an article on the growth of ecotourism in Costa Rica, journalist Martha Honey writes that in 1992, the U.S. Adventure Travel Society identified Costa Rica as the top ecotourism destination worldwide. The following year, income from tourism surpassed that of coffee and bananas as the top foreign exchange earner (40).

5. For more on this history, see Paul F. Steinberg's *Environmental Leadership in Developing Countries* (2001). In this book, Steinberg offers a thoroughly researched account of the complex web of social, political, and scientific relationships that led to the creation of policies of environmental protection in Costa Rica.

6. One such convention was the Convention on the Trade in Endangered Species (CITIES) ratified by Costa Rica in 1974.

7. See Nelson, particularly the chapter on "The Economy."

8. I witnessed just such an incident with a colleague in January 2002, during a visit to Monteverde, Costa Rica. We were walking with a guide and students along a trail in a preserve, when upon rounding a bend, we happened upon a large group of tourists staring intently through the forest toward a solitary quetzal perched several meters away.

9. *Marianismo* is the counterpart to machismo, that is, it is a prescribed code of conduct for women. Machismo emphasizes stereotypically masculine traits of protectiveness, assertiveness, strength, and sexual aggression. *Marianismo* comes from the name María and holds up the Virgin Mary as the womanly ideal: pure, maternal, self-sacrificing, and long-suffering.

10. For more on Quince Duncan's representation of the struggle of people of West Indian ethnicity for social inclusion in Costa Rica, see Dorothy Mosby's book *Place, Language, and Identity in Afro-Costa Rican Literature*.

11. Foundational myths are stories or images that encode a mental image of the affinity among members of a given nation. For more on the notion of the nation as an "imagined community," see Benedict Anderson's *Imagined Communities*. For more on the idea of foundational fictions in Latin America, see Doris Sommer's *Foundational Fictions: The National Romances of Latin America*.

12. In 1970, just a few years prior to the publication of Gutiérrez's novel, the twin concerns about environment and sovereignty coalesced in protests against a contract awarded to the North American aluminum company Alcoa to establish a bauxite processing plant (Steinberg 62–63).

13. Dr. Vining Dunlap, a United Fruit scientist, discovered in the 1940s that "flood fallowing" was an innovative means of reclaiming land infected with Panama disease for banana cultivation. His method consisted of converting infected farms into shallow lakes for a period of three to eighteen months, then draining and replanting them (Soluri 167–68).

14. See García Canclini's *Culturas híbridas: estrategias para entrar y salir de la modernidad*.

15. With the tidal wave, Lobo accelerates in fiction a fact of the Talamanca coast in the twentieth century: the encroaching Caribbean. According to Paula Palmer, longtime residents recall that "shorelines extended out as much as two hundred meters beyond today's beaches" (65).

16. Ethnobotanical studies validate Rossi's representation. For example, García Serrano and Del Monte write that "religious beliefs and myths influence the use of plants, and observance of ancestral rules and restrictions, which govern especially the use of wild species, . . . contributes to the sustainable use of plant resources" (67). They go on to note that both botanical knowledge and biodiversity are disappearing, and it is hard to determine which is the consequence and which is the cause (67).

17. Yemanyá is an orisha, or deity, associated with the sea in the African Yoruba religion.

18. This gesture outward to readers is significant, and not just as a call to consciousness. According to Steinberg, "the political knowledge, skills, and contacts needed to win [environmental] struggles require a long-term, in-country presence that few foreign advocates or international organizations possess" (7). The novel is instructive to readers outside of Costa Rica in that it depicts the formidable resources, particularly social ones, required to protect the environment, as well as those summoned on the side of unsustainable development. Failure on the part of well-meaning outsiders to appreciate the social and cultural dynamics of a country can produce unsuccessful outcomes and even prove counterproductive. Daniela's last act in the novel beckons the reader in to understand this world (the human and nonhuman one alike) as she does.

Chapter 4. Futuristic Narratives and the Crisis of Place

1. Biologist Edward O. Wilson published *Biophilia* in 1984 and proposed that biophilia marks human beings, that is, they have "the innate tendency to focus on life and lifelike processes" (1). Subsequent publications, like *The Biophilia Hypothesis* (1993) by Stephen R. Kellert and E. O. Wilson, contain arguments about biophilia by proponents and critics of the concept.

2. Interestingly, the cover art on Lawrence Buell's book *The Future of Environmental Criticism* features a picture of Caracas, Venezuela, a megalopolis of profound income disparity.

3. Aridjis' choice of Ariel as a name is significant and not just for the Shakespearean allusion. *Ariel* (1900), by José Enrique Rodó, is one of the most important essays in Latin American "modernismo." It extols the spiritual and creative wealth of Latin America, while contrasting this against the pragmatism of the United States.

4. Aridjis noted that he set his futuristic novel in 2027 to conform to the Aztec ritual calendar: "Yo escogí una fecha ritual del calendario del México Antiguo. Según esa referencia, en el año 2027 tendrá lugar el próximo fuego nuevo de los aztecas. Fijo el año 2027 como la fecha del fin del Quinto Sol. Cabe recordar que vivimos bajo el signo del Quinto Sol, Olin Tonatiuh, el cuatro movimiento o Sol que camina hacia su muerte. El Quinto Sol va a terminar por terremotos" (J. Castro) [I chose a ritual date from the ancient Mexican calendar. According to this reference, 2027 begins the age of the next sun. I made 2027 the date of the end of the Fifth Sun. It's worth remembering that we live under the Fifth Sun, Olin Tonatiuh, the fourth movement or Sun that is marching toward its death. The Fifth Sun will end in earthquakes].

5. Belli first published *Waslala* in 1996 with Anamá Ediciones in Nicaragua. When she published it again a decade later, this time with Seix Barral, she made numerous editorial changes to the text. I have quoted from the 2006 edition because the first edition, with a run of only three thousand copies, is hard to find.

6. Arne Naess, who coined the term "deep ecology," associates the concept with an awareness of humans as part of a larger natural community. "Deep ecology" contrasts with "shallow ecology," which focuses more upon front-line activism in the face of individual threats to the environment (Naess 95–100). Recently, though, authors like William Chaloupka have criticized the deep/shallow dichotomy, especially as it has been used in debates about the constructedness of nature. Chaloupka argues that "green intellectuals cause no end of needless trouble for their movement when they act on their confidence (that they have developed a governing, masterful way of understanding the world). Greens arrogantly proclaim that they can discern what is 'deep' (as in 'deep ecology') and what is shallow—and that this distinction matters. Their possession of a perspective identified with nature leads them to casually identify a 'biocentric' position from which they can denounce all that is, conversely 'anthropocentric'" (31).

7. According to Leonel Delgado Aburto, it is not just geographically significant but also culturally imposing: "El río San Juan ha mitificado su geografía gracias a las buenas o malas intenciones de las potencies extranjeras y las élites nacionales" (1) [The Río San Juan has mythologized the geography thanks to the good or bad intentions of foreign powers and national elites]. In Belli's work the river has multiple significations: source of inspiration for poets, anchor for collective memory, and vehicle for foreign penetration in a rich territory.

8. In a note at the end of the text, Belli points out that the Cesium scene in *Waslala* is drawn from the pages of Latin American history, specifically a 1987 event in which several people died in Brazil. Among those who died was a young girl who had celebrated her birthday amid the radioactive glow of Cesium from a container found in a dump.

9. See, for example, *Seven Trees against the Dying Light*, a bilingual edition of Spanish poetry by Nicaraguan Pablo Antonio Cuadra.

Works Cited

The 11th Hour. Dir. Nadia Conners and Leila Conners Petersen. Written by Nadia Conners and Leonardo DiCaprio. Warner Bros., 2007. Film.

Adamson, Joni, Mei Mei Evans, and Rachel Stein. "Introduction: Environmental Justice Politics, Poetics, and Pedagogy." Adamson, Evans, and Stein 3–14. Print.

Adamson, Joni, Mei Mei Evans, and Rachel Stein. *The Environmental Justice Reader.* Tucson, AZ: University of Arizona Press, 2002. Print.

Anderson, Benedict. *Imagined Communities: Reflections on the Origin and Spread of Nationalism,* 2nd ed. New York: Verso, 1991. Print.

Arellano, Jorge Eduardo, ed. Prólogo. *Pablo Antonio Cuadra: Poesía selecta.* Caracas: Ayacucho, 1991. IX-XLVIII. Print.

Aridjis, Homero. *¿En quién piensas cuando haces el amor?* México, D.F.: Alfaguara, 1995. Print.

Augenbraum, Harold. "Review of Fordlandia." *Library Journal* 125.15 (15 Sept. 2000): 114. InfoTrac. Web. 26 June 2006.

Averbach, Márgara. "Encuentro indio-blanco a principios del siglo: las dos puntas de América, Fuegia y Mean Spirit." *Primeras Jornadas Internacionales de Literatura Argentina Comparatística.* Ed. Teresita Frugoni de Fritzsche. Buenos Aires: Facultad de Filosofía y Letras, Universidad de Buenos Aires, 1995. 61–72. Print.

Balée, William L. *Footprints of the Forest: Ka'apor Ethnobotany—the Historical Ecology of Plant Utilization by an Amazonian People.* New York: Columbia University Press, 1994. Print.

Barbosa, Luiz C. "The People of the Forest against International Capitalism: Systemic and Anti-Systemic Forces in the Battle for the Preservation of the Brazilian Amazon Rainforest." *Sociological Perspectives* 39.2 (Summer 1996): 317–31. JSTOR. Web. 1 Sept. 2009.

Barbas-Rhoden, Laura. "Ecology, Coloniality, Modernity: Argentine Fictions of Tierra del Fuego." *Mosaic* 41.1 (2008): 1–18. Print.

Bate, Jonathan. *Song of the Earth.* Cambridge: Harvard University Press, 2000. Print.

Belgrano Rawson, Eduardo. *Fuegia*, 6th ed. Buenos Aires: Editorial Sudamericana, 1997. Print.

Belli, Gioconda. *Waslala*. Barcelona, Spain: Seix Barral, 2006.

———. *El país bajo mi piel*. Barcelona, Spain: Plaza y Janés, 2001.

Binns, Niall. "Landscapes of Hope and Destruction: Ecological Poetry in Spanish America." *The ISLE Reader. Ecocriticism, 1993–2003*. Ed. Michael P. Branch and Scott Slovic. Athens, GA: University of Georgia Press, 2003. 124–39. Print.

Bolay, Jean-Claude, Yves Pedrazzini, Adriana Rabinovich, Andrea Catenazzi, Carlos García Pleyán. "Urban Environment, Spatial Fragmentation and Social Segregation in Latin America: Where Does Innovation Lie?" *Habitat International* 29 (2005): 627–45. Academic Search Premier. Web. 21 Jan. 2010.

Booth, Wayne. *The Company We Keep: An Ethics of Fiction*. Berkeley: University of California Press, 1988. Print.

Bortz, Jeffrey, and Marcos Aguila. "Earning a Living. A History of Real Wage Studies in Twentieth-Century Mexico." *Latin American Research Review* 41.2 (2006): 112–38. Print.

Boyce, James K. "Inequality as a Cause of Environmental Degradation." 1994. Published Studies ps1, Political Economy Research Institute. University of Massachusetts at Amherst. Web. 24 Feb. 2009.

Brooke, James. "Tourism Site Springs from a Nuclear Horror Story." *New York Times*. 3 May 1995: A6. ProQuest Historical Newspapers. Web. 20 Jan. 2010.

Buell, Lawrence. *The Environmental Imagination: Thoreau, Nature Writing, and the Formation of American Culture*. Cambridge: Belknap Press of Harvard University Press, 1995. Print.

———. *The Future of Environmental Criticism: Environmental Crisis and the Literary Imagination*. Malden, MA: Blackwell, 2005. Print.

———. "Toxic Discourse." *Critical Inquiry* 24.3 (1998): 639–65. Print.

Bulmer-Thomas, Victor. *The Economic History of Latin America since Independence*. New York: Cambridge University Press, 1994. Print.

Camara, Mario. "Entrevista a Eduardo Sguiglia." Leedor.com. N.p. 22 Feb. 2001. Web. 26 June 2006.

Campanario, Sebastián. "Cómo escriben los economistas." 5 Oct. 2003. Clarín.com. Web. 20 Jan 2010.

Candido, Antonio. "Literature and Underdevelopment." Del Sarto, Ríos, and Trigo 35–57. Print.

Caribbean Marine Protected Areas Database. "Gandoca-Manzanillo." Caribbean Marine Protected Areas Management. Caribbean Marine Protected Areas Network and Forum. n.d. Web. 26 Feb 2009.

Carpenter, Frank G. *South America: Social, Industrial, and Political*. Boston: George M. Smith and Company, 1900. Print.

Casini, Silvia. "Luis Sepúlveda: un viaje express al corazón de la Patagonia." *Alpha* 20 (2004): 103–20. Print.

Castro, José Alberto. "Fuentes, Aridjis y Agustín encabezan un grupo heterogéneo de narradores futuristas." *InfoLatina*. 10 Aug. 1997. LexisNexis Academic. Web. 7 Oct. 2005.

Castro, Marcela, and Silvia Jurovietzky. "Una escritora de 'perfil bajo': entrevista a Libertad Demitrópulos." *Feminaria literaria* 8.13 (1997): 66–69. Print.

Chaloupka, William. "Jagged Terrain: Cronon, Soulé, and the Struggle over Nature and Deconstruction in Environmental Theory." *Strategies* 13.1 (2000): 23–38. Academic Search Premier. Web. 20 Jan. 2010.

Chapman, Anne. *Drama and Power in a Hunting Society: The Selk'nam of Tierra del Fuego.* Cambridge: Cambridge University Press, 1982. Print.

Cheadle, Norman. "Rememorando la historia decimonónica desde *La tierra del fuego* (1998) de Sylvia Iparraguirre." *Celebración de la creación de escritoras hispanas en las Américas.* Ed. Lady Rojas-Trempe and Catharina Vallejo. Ottawa, Canada: GIROL Books, 2000. 81–91. Print.

Cleary, David. "Towards an Environmental History of the Amazon: From Prehistory to the Nineteenth Century." *Latin American Research Review* 36.2 (2001): 65–96. Print.

Collin, Robin Morris. "The Apocalyptic Vision, Environmentalism, and a Wider Embrace." *ISLE* 13.1 (2006): 1–11. Print.

Cuadra, Pablo Antonio. *Seven Trees against the Dying Light: A Bilingual Edition.* Trans. Greg Simon and Steven F. White. Evanston: Northwestern University Press, 2007. Print.

Darwin, Charles. *Voyage of the Beagle.* New York: Collier, 1909. Print.

de Fays, Hélène. "Neo-Luddism in a Mexican Novel: *¿En quién piensas cuando haces el amor?* by Homero Aridjis." *CiberLetras* 4 (2001). Web. 19 Jan. 2010.

Delgado Aburto, Leonel. "En busca de la genealogía de Cifar." *Istmo* (enero–junio 2001). Web. 30 Mar. 2008.

DeLoughrey, Elizabeth M., George B. Handley, and Renee K. Gosson, eds. *Caribbean Literature and the Environment: Between Nature and Culture.* Charlottesville: University of Virginia Press, 2005. Print.

Del Sarto, Ana, Alicia Ríos, and Abril Trigo, eds. *The Latin American Cultural Studies Reader.* Durham: Duke University Press, 2004. Print.

Demitrópulos, Libertad. *Un piano en Bahía Desolación.* Buenos Aires: Ediciones Braga, 1994. Print.

———. *Río de las congojas.* Buenos Aires: Editorial Sudamericana, 1981. Print.

———. *River of Sorrows.* Trans. Mary G. Berg. Buffalo: White Pine, 2000.

Dempsey, Mary A. "Fordlandia." *Michigan History* (July/August 1994). Web. 30 Aug. 2005.

Denevan, William. "The Pristine Myth: The Landscape of the Americas in 1492." *Annals of the Association of American Geographers* 82.3 (1992): 369–85. Academic Search Premier. Web. 10 Jan 2006.

De Souza, Aguiar, Marco Antonio Marcos Arruda, Parsifal Flores, and Terrie Groth. "Economic Dictatorship versus Democracy in Brazil." *Latin American Perspectives* 11.1 (1984): 13–25. JSTOR. Web. 1 Sept. 2009.

Despres, Leo A. *Manaus.* Albany: SUNY Press, 1991. Print.

DiAntonio, Robert E. "The Aesthetics of the Absurd in *Galvez, Imperador do Acre*: The Novel as Comic Opera." *Hispania* 70.2 (1987): 265–70. JSTOR. Web. 1 Aug. 2009.

Don, Rubén. "De aire, agua y tierra . . . Diálogo con el mexicano Homero Aridjis." *LibrUSA.* Aug. 1998. Web. 25 July 2005.

Dosal, Paul J. *Doing Business with Dictators: A Political History of United Fruit in Guatemala, 1899–1944*. Wilmington: Scholarly Resources, 1993. Print.

Dussel, Enrique. "Europe, Modernity, and Eurocentrism." *Nepantla* 1.3 (2000): 465–78. Print.

———. "World-System and 'Trans'-Modernity." *Nepantla* 3.2 (2002): 221–44. Print.

"Escribir de Oído." Literama.net. Feb. 2002. Web. 20 Feb. 2003.

Faiz, Asif, Surhid Gautam, and Emaad Burki. "Air Pollution from Motor Vehicles: Issues and Options for Latin American Countries." *The Science of the Total Environment* 169 (1995): 303–10. ScienceDirect. Web. 22 Jan. 2010.

Feeney, Patricia. "Environmental Reform in Brazil: Advances and Reversals." *Development in Practice* 2.1 (1992): 3–11. JSTOR. Web. 1 Sept. 2009.

French, Jennifer. *Nature, Neo-colonialism and the South American Regional Writers*. Hanover: Dartmouth College Press, 2005. Print.

Gaard, Greta. "Strategies for a Cross-Cultural Ecofeminist Ethics: Interrogating Tradition, Preserving Nature." *The Bucknell Review* 44.1 (2000): 82–101. MLA International Bibliography. Web. 22 Jan. 2010.

Gaard, Greta, and Patrick D. Murphy. "A Dialogue on the Role and Place of Literary Criticism within Ecofeminsim." *Isle: Interdisciplinary Studies in Literature and Environment* 3.1 (1996): 1–6. Print.

García Canclini, Néstor. *Culturas híbridas: estrategias para entrar y salir de la modernidad*. Buenos Aires: Paidós, 2001. Print.

———. "Cultural Studies from the 1980s to the 1990s: Anthropological and Sociological Perspectives in Latin America." Del Sarto, Ríos, and Trigo 329–46. Print.

García Serrano, Carlos Ramos, and Juan Pablo Del Monte. "The Use of Tropical Forest (Agrosystems and Wild Plant Harvesting) as a Source of Food in the Bribri and Cabecar Cultures in the Caribbean Coast of Costa Rica." *Economic Botany* 58.1 (2004): 58–71. Expanded Academic ASAP. Web. 26 Feb. 2009.

Gauld, Charles A. *The Last Titan: Percival Farquhar*. Stanford: Institute of Hispanic American and Luso-Brazilian Studies, Stanford University, 1964. Print.

Genette, Gérard. *Narrative Discourse*. Trans. Jane E. Lewin. Ithaca: Cornell University Press, 1980.

Gerlach, Allen. *Indians, Oil, and Politics: A Recent History of Ecuador*. Wilmington: Scholarly Resources, 2003. Print.

Gilbert, Helen. "Belated Journeys: Ecotourism as a Style of Travel Performance." In *Transit: Travel, Text, Empire*. Ed. Helen Gilbert and Anna Johnston. New York: Lang, 2002. 257–74.

Girardet, Herbert. "The Ethno-Ecologist Revealed." *The Ecologist* 31.4 (2001): 57. Expanded Academic ASAP. Web. 26 Feb. 2009.

Golden, Tim. "Mexico's Capital Is Felled By Smog." *New York Times* 18 Mar. 1992: A5. ProQuest Historical Newspapers. Web. 22 Jan. 2010.

Gomides, Camilo, and Joseph Henry Vogel. "An Ecocritical Analysis of *The Old Man Who Read Love Stories* by Luis Sepúlveda." *Ometeca* 9 (2005). Web. 25 Feb. 2009.

González, Daniuska. "Viaje a la narrative de Luis Sepúlveda. Escribir la ecología: la nueva mirada del escritor viajero." *Revista de literatura hispanoamericana* 48 (2004): 7–24. Print.

Goodrich, Diana Sorensen. *Facundo and the Construction of Argentine Culture.* Austin: University of Texas Press, 1996. Print.

Graham-Yool, Andrew. "Light at the End of the Tunnel." *Antioch Review* 52.4 (1994): 566–80. Academic Search Premier. Web. 20 Jan. 2010.

Guha, Ramachandra. *Environmentalism. A Global History.* New York: Longman, 2000. Print.

Guha, Ramachandra, and J. Martínez-Alier. *Varieties of Enviromentalism: Essays North and South.* London: Earthscan Publications, 1997. Print.

Gutiérrez, Joaquín. *Murámonos, Federico.* San José: Editorial Costa Rica, 1973. Print.

Hazlewood, Nick. *Savage: The Life and Times of Jemmy Button.* Thomas Dunne Books, 2001. Print.

Heckenberger, Michael J. *The Ecology of Power: Culture, Place, and Personhood in the Southern Amazon, A.D. 1000–2000.* New York: Routledge, 2005. Print.

Hilton, Ronald. Foreword. *The Last Titan: Percival Farquhar.* Charles A. Gauld. Stanford, CA: Institute of Hispanic American and Luso-Brazilian Studies, Stanford University, 1964. x–xi. Print.

Honey, Martha. "Giving a Grade to Costa Rica's Green Tourism." *NACLA Report on the Americas* 36.6 (2003): 39–47. Expanded Academic ASAP. Web. 25 Feb. 2009.

Iparraguirre, Sylvia. *La tierra del fuego.* Buenos Aires: Alfaguara, 1998. Print.

———. *Tierra del fuego.* Trans. Hardie St. Martin. Willimantic, CT: Curbstone Press, 2000. Print.

Iyengar, G. Venkatesh, and Padmanabhan P. Nair. "Global Outlook on Nutrition and the Environment: Meeting the Challenges of the Next Millennium." *The Science of the Total Environment* 249 (2000): 331–46. ScienceDirect. Web. 22 Jan. 2010.

Kearns, Sofía. "Otra cara de Costa Rica a través de un testimonio ecofeminista." *Hispanic Journal* 19.2 (1998): 313–39. Print.

Kellert, Stephen R. *The Value of Life: Biological Diversity and Human Society.* Washington, D.C.: Island Press, 1996.

Kellert, Stephen R., and Edward O. Wilson. *The Biophilia Hypothesis.* Washington, D.C: Island Press, 1993. Print.

Kokotovic, Misha. "After the Revolution: Central American Narrative in the Age of Neoliberalism." *A contracorriente* 1.1 (2003): 19–50. MLA International Bibliography. Web. 22 Jan. 2010.

Kricher, John. *A Neotropical Companion: An Introduction to the Animals, Plants, and Ecosystems of the New World Tropics,* 2nd ed. Princeton: Princeton University Press, 1997. Print.

Lahar, Stephanie. "Roots: Rejoining Natural and Social History." *Ecofeminism: Women, Animals, Nature.* Ed. Greta Gaard. Philadelphia: Temple University Press, 1993. 91–117. Print.

Lander, Edgardo. "Eurocentrism and Colonialism in Latin American Social Thought." *Nepantla* 1.3 (2000): 519–32. Print.

Langley, Lester D., and Thomas Schoonover. *The Banana Men: American Mercenaries and Entrepreneurs in Central America, 1880–1930*. Lexington, KY: University Press of Kentucky, 1995. Print.

León-Portilla, Miguel. *The Aztec Image of Self and Society*. Trans. Charles E. Bowden and J. Jorge Klor de Alva. Salt Lake City: University of Utah Press, 1992. Print.

Lobo, Tatiana. *Calypso*. San Pedro: Farben Grupo Editorial Norma, 1996. Print.

Locke, John. *Second Treatise of Government*. Ed. C. B. Macpherson. Cambridge: Hackett, 1980. Print.

Longhini, Nora. "Las voces olvidadas en *Fuegia*, de Eduardo Belgrano Rawson." *Primeras Jornadas Internacionales de Literatura Argentina Comparatística*. Ed. Teresita Frugoni de Fritzsche. Buenos Aires: Facultad de Filosofía y Letras, Universidad de Buenos Aires, 1995. 73–80. Print.

López, Miguel. "Pensar la nación a través del Apocalipsis ecológico en dos novelas distópicas de Homero Aridjis." *La luz queda en el aire: estudios internacionales en torno a Homero Aridjis*. Ed. Thomas Stauder. Bismarck, Ger.: Vervuert, 2005. 173–86. Print.

Love, Glen A. *Practical Ecocriticism: Literature, Biology, and the Environment*. Charlottesville: University of Virginia Press, 2003. Print.

———. "Revaluing Nature: Toward an Ecological Criticism." *The Ecocriticism Reader: Landmarks in Literary Ecology*. Eds. Cheryll Glotfelty and Harold Fromm. Athens, GA: University of Georgia Press, 1996. 225–40. Print.

Lynch, Kathryn R. "Environmental Education and Conservation in Southern Ecuador: Constructing an Engaged Political Ecology Approach." Diss. University of Florida, 2001. Print.

Malaver Rodríguez, Rodrigo. "La selva imaginada: una relectura crítica de *Un viejo que leía novelas de amor* de Luis Sepúlveda." *Cuadernos de literatura* 7.13–14 (2001): 31–44. Print.

"Manifiesto de las comunidades indígenas afectadas por el eventual Proyecto Hidroeléctrico Boruca." *Ambientico* 91 (2001): 22. Print.

Marcos, Subcomandante. "The Fourth World War Has Begun." Trans. Ed Emery. *Nepantla* 2.3 (2001): 559–72. Print.

Marks, Robert B. *The Origins of the Modern World: A Global and Ecological Narrative*. Lanham, MD: Rowman and Littlefield, 2002. Print.

Marquardt, Steve. "Pesticides, Parakeets, and Unions in the Costa Rican Banana Industry, 1938–1962." *Latin American Research Review* 37.2 (2001): 3–36. Print.

Masiello, Francine. *Between Civilization and Barbarism: Women, Nation, and Literary Culture in Modern Argentina*. Lincoln: University of Nebraska Press, 1992. Print.

Mathieu, Corina S. "Fuegia: Crónica de una marginación." *Alba de América* 14.26–27 (1996): 147–54. Print.

Merchant, Carolyn. *Ecological Revolutions: Nature, Gender and Science in New England*. Chapel Hill: University of North Carolina Press, 1989.

Mignolo, Walter. *Local Histories/Global Designs: Coloniality, Subaltern Knowledges, and Border Thinking*. Princeton: Princeton University Press, 2000. Print.

Mora, Sonia Marta. "Joaquín Gutiérrez y la culminación de la novela costarricense." *Revista iberoamericana* 53.138–39 (1987): 245–63. Print.

———."*Murámonos, Federico* o la insinuación de la esperanza." *Káñina* 12.2 (1988): 23–27. Print.

Moran, Emilio F. "Deforestation and Land Use in the Brazilian Amazon." *Human Ecology* 21.1 (1993): 1–21. JSTOR. Web. 1 Sept. 2009.

Mosby, Dorothy. *Place, Language, and Identity in Afro-Costa Rican Literature.* Columbia: University of Missouri Press, 2003. Print.

Mosley, Stephen. "Common Ground: Integrating Social and Environmental History." *Journal of Social History.* 39.3 (2006): 915–34. JSTOR. Web. 22 Jan. 2010.

"Murder of Missionaries in Patagonia." *New York Times.* 14 June 1860: 5. ProQuest Historical Newspapers. Web. 22 Jan. 2010.

Murphy, Patrick D. Introduction. *Hispanic Journal* 19.2 (1998): 203–8. Print.

Nabhan, Gary Paul. *Cultures of Habitat: On Nature, Culture, and Story.* Washington, D.C.: Counterpoint, 1997. Print.

Naess, Arne. "The Shallow and the Deep Long-Range Ecology Movement: A Summary." *Inquiry* 16 (1973): 95–100. Print.

Nature Conservancy 56.2 (2006). Print.

Nelson, Harold D. *Costa Rica: A Country Study.* Washington, D.C.: U.S. GPO, 1983. Print.

Nevins, Allan, and Frank Ernest Hill. *Ford: Expansion and Challenge, 1915–1933.* New York: Scribner's, 1957. Print.

Nichols, Peter. *Evolution's Captain.* New York: HarperCollins, 2003. Print.

Nygren, Anja. "Struggle over Meanings: Reconstruction of Indigenous Mythology, Cultural Identity, and Social Representation." *Ethnohistory* 45.1 (1998): 31–63. JSTOR. Web. 22 Jan. 2010.

O'Riordan, Timothy. "On Justice, Sustainability, and Democracy." *Environment* 47.6 (2005): 0+. Academic Search Premier. Web. 16 June 2006.

Palmer, Paula. *What Happen: A Folk History of Costa Rica's Talamanca Coast,* 1st ed., rev. San José: Editorama, 1993. Print.

Pérez Martin, Norma. "Dos novelas 'patagónicas' en la narrativa argentina actual: *Fuegia* y *El Rey de la Patagonia*." *Alba de América: Revista Literaria* 15.28–29 (1997): 142–47. Print.

Perkowska-Alvarez, Magdalena. "Memorias de un mundo perdido: rememoración y evocación en *Fuegia* de Eduardo Belgano Rawson." *Mémoire et culture en Amérique Latine.* Ed. Christian Guidicelli. Paris: Presses de la Sorbonne Nouvelle, 2002. 83–90. Print.

Plumwood, Val. *Feminism and the Mastery of Nature.* London: Routledge, 1993. Print.

Posey, Darrell A. "Protecting Indigenous Peoples' Rights to Biodiversity." *Environment* 38.6 (1996): 6–9, 37–45. Print.

Pratt, Mary Louise. *Imperial Eyes: Travel Writing and Transculturation.* London: Routledge, 1992. Print.

"Proposal for the Design of a Biosphere Reserve in the South of Ecuador." *UNESCO.* n.d. Web. 19 Aug. 06.

Quijano, Aníbal. "Modernity, Identity, and Utopia in Latin America." *boundary* 2 20.3 (1993): 140–55. JSTOR. Web. 22 Jan. 2010.

Quijano, Aníbal, and Immanuel Wallerstein. "Americanity as a Concept, or the Americas in the Modern World-System." *The International Social Science Journal* 44.4 (1992): 549–58. Print.

Raffles, Hugh, and Antoinette M.G.A. WinklerPrins, "Further Reflections on Amazonian Environmental History: Transformations of Rivers and Streams." *Latin American Research Review* 38.3 (2003): 165–87. Print.

Reed, T. V. "Toward an Environmental Justice Ecocriticism." Adamson, Evans, and Stein 145–62. Print.

Ríos, Alicia. "Forerunners." Del Sarto, Ríos, and Trigo 15–34. Print.

Rivas, Gabriel. "Costa Rica libre de exploración y explotación petroleras." *Ambientico* 85 (2000): 6–7. Print.

Roberts, Bryan R., and Alejandro Portes. "Coping with the Free Market City. Collective Action in Six Latin American Cities at the End of the Twentieth Century." *Latin American Research Review* 41.2 (2006): 57–83. Print.

Roberts, J. Timmons, and Nikki Demetria Thanos. *Trouble in Paradise: Globalization and Environmental Crises in Latin America*. New York: Routledge, 2003. Print.

Rock, David. *Argentina, 1516–1987*. Berkeley: University of California Press, 1987. Print.

Rodó, José Enrique. *Ariel*. Trans. Margaret Sayers Peden. Austin: University of Texas Press, 1988. Print.

Roffé, Reina. "Entrevista a Sylvia Iparraguirre." *Cuadernos hispanoamericanos* 603 (2000): 99–106. Print.

Rohter, Larry. "Amazon Writers Breaking Out of the Mold." *International Herald Tribune*. 26 Sept. 2007: Feature 12. LexisNexis. Web. 18 Nov. 2008.

Rosendale, Steven. Introduction. *The Greening of Literary Scholarship: Literature, Theory, and the Environment*. Iowa City: University of Iowa Press, 2002. Print.

Rossi, Anacristina. *La loca de Gandoca*. San José, Costa Rica: EDUCA, 1995. Print.

Rotker, Susana. *Captive Women: Oblivion and Memory in Argentina*. Minneapolis: University of Minnesota Press, 2002. Print.

Russell, Dick. "Homero Aridjis y la ecología." Trans. Amina Roth. *La luz queda en el aire: estudios internacionales en torno a Homero Aridjis*. Ed. Thomas Stauder. Bismarck, Ger.: Vervuert, 2005. 65–81. Print.

Sachs, Aaron. *Eco-Justice: Linking Human Rights and the Environment*. Washington, D.C.: Worldwatch Institute, 1995. Print.

Sachs, Jeffrey. "Does Conservation Matter to the Poor?" *Nature Conservancy* 56.2 (2006): 28. Print.

Sarmiento, Domingo Faustino. *Facundo: Civilization and Barbarism*. Trans. Kathleen Ross. Berkeley: University of California Press, 2003. Print.

———. *Facundo*. Buenos Aires: Centro Editor de América Latina, 1967. Print.

"Sea and Ship News: Voyage of the Clipper Ship *Great Republic*." *New York Times* 6 July 1858: 3. ProQuest Historical Newspapers. Web. 22 Jan. 2010.

Sepúlveda, Luis. *The Old Man Who Read Love Stories*. Trans. Peter Bush. Orlando: Harcourt Brace, 1993. Print.

———. *Un viejo que leía novelas de amor*. New York: Penguin Books, 1998. Print.

Sguiglia, Eduardo. *Fordlandia*. Buenos Aires: Grupo Editorial Norma, 1997. Print.

———. *Fordlandia*. Trans. Patricia J. Duncan. New York: St. Martin's Press, 2000. Print.

Sheck, Ree Strange. *Costa Rica: A Natural Destination*. Santa Fe: Muir Publications, 1990. Print.

Sheppard, James W. "Book Review of *Environmentalism: A Global History* by Ramachandra Guha." *Ethics and the Environment* 8.2 (2003): 132–39. Print.

Shipton, Eric. *Tierra del Fuego: The Fatal Lodestone*. London: Knight, 1973. Print.

Slater, Candace. *Entangled Edens*. Berkeley: University of California Press, 2002. Print.

Slovic, Scott. "Ecocriticism with or without Narrative: The Language of Conscious Experience versus the Language of Freefall." *Narrative Scholarship: Storytelling in Ecocriticism*. ASLE Ecocritical Library. n.d. Web. 4 Sept. 2008.

Smith, Nigel J. H. "Colonization Lessons from a Tropical Forest." *Science* 214.4522 (13 Nov. 1981): 755–61. JSTOR. Web. 1 Sept. 2009.

Soluri, John. *Banana Cultures: Agriculture, Consumption & Environmental Change in Honduras and the United States*. Austin: University of Texas Press, 2005. Print.

Sommer, Doris. "For Love and Money: Of Potboilers and Precautions." *PMLA* 116.2 (2001): 380–91. Print.

——. *Foundational Fictions: The National Romances of Latin America*. Berkeley: University of California Press, 1991. Print.

Souza, Márcio. *Mad Maria*, 2nd ed. Rio de Janeiro: Civilização Brasileira, 1980. Print.

——. *Mad Maria*. Trans. Thomas Colchie. New York: Avon, 1985. Print.

Stavans, Ilan. "World Literature in Review: Spanish." Review of *Nombre del torero* by Luis Sepúlveda. *World Literature Today* 70.2 (1996): 369. Academic Search Premier. Web. 23 June 2006.

Steinberg, Paul F. *Environmental Leadership in Developing Countries: Transnational Relations and Biodiversity Policy in Costa Rica and Bolivia*. Boston: MIT Press, 2001. Print.

Summerhill, William R. "Big Social Savings in a Small Laggard Economy: Railroad-Led Growth in Brazil." *Journal of Economic History* 65.1 (2005): 72–102. Web. 19 Jan. 2009. JSTOR. Web. 1 Sept. 2009.

Swinton, Scott M., Germán Escobar, and Thomas Reardon. "Poverty and Environment in Latin America: Concepts, Evidence and Policy Implications." *World Development* 31.11 (2003): 1865–72. Academic Search Premier. Web. 22 Jan. 2010.

Tedlock, Dennis, ed. and trans. *Popol Vuh: The Mayan Book of the Dawn of Life*. New York: Simon and Schuster, 1985. Print.

Thompson, Harry. *This Thing of Darkness*. London: Headline Book Publishing, 2005. Print.

Trejos, Alonso. *Illustrated Geography of Costa Rica*, 2nd ed. Trans. Harry Spencer. San José, Costa Rica: Trejos Hermanos Sucesores, 1996. Print.

Tucker, Richard P. *Insatiable Appetite: The United States and the Ecological Degradation of the Tropical World*. Berkeley: University of California Press, 2000. Print.

Urbina, Nicasio. "La literatura centroamericana." *Istmo* 3 (2001). Web. 16 Feb. 2009.

Williams, Robert G. *States and Social Evolution: Coffee and the Rise of National Governments in Central America*. Chapel Hill: University of North Carolina Press, 1994. Print.

Wilson, David J. *Indigenous South Americans of the Past and Present: An Ecological Perspective*. Boulder: Westview Press, 1999. Print.

Wilson, Edward O. *Biophilia*. Cambridge: Harvard University Press, 1984. Print.

Winseck, David R., and Robert M. Pike. *Communication and Empire: Media, Markets, and Globalization, 1860–1930*. Durham: Duke University Press, 2007. Print.

Wylynko, David. "The Ecological Ethic: Sustainability, Ecophilosophy, and Literary Criticism." *Canadian Poetry* 42 (1998): 123–39. MLA International Bibliography. Web. 22 Jan. 2010.

Yardley, Jonathan. "The *Post's* Book Critic Selects His Favorites of the Year." *Washington Post* 3 Dec. 2000: X02. LexisNexis. Web. 18 Nov. 2008.

Index

Laura Barbas-Rhoden is an associate professor of foreign languages at Wofford College in Spartanburg, South Carolina. She is the author of *Writing Women in Central America* (Ohio University Press, 2003).